1944

GRANGE FAIR

CENTRE HALL, PENNA.

GENERAL ADM

SHOW YOUR

To be readmitted to the grounds, present both ticket stub and Pass-out check for the day.

Admission	67c
Federal Tax	13c
Total	80c

ONE ADMISSION

GRANGE HALL

Centre County
Grange Encampment and Fair

"A History of Master Rhone's Pic-nic"

Norman K. Lathbury

August 27, 1999

To Jean Strouse,

Best Wishes and enjoy the fair.

[illegible]

Centre County Grange Fair and Encampment "A History of Master Rhone's Pic-nic"

Published by:

P.O. Box 345
Centre Hall, Pennsylvania 16828
(814) 364-2000 • (800) 200-5375 • Fax (814) 364-9999
in cooperation with:
The Centre County Grange Encampment and Fair History Book Committee.

Library of Congress catalog card number:
99-65604

ISBN number: 0-9674051-0-6

Dedication

This book is dedicated to all the people, who for generations, have come from near and far to make this Fair a success. It is also dedicated to Centre County Grangers and volunteers, past and present, who have given freely of their time and talent to make this Fair possible. Their remarkable perseverance and determination to continue the founder's legacy, is exemplified in the 125th Anniversary of the Grange Encampment and Centre County Fair.

Forward

Writing and publishing the history of the Grange Encampment and Centre County Fair has proven to be an extraordinary challenge. We are extremely grateful to all who took the time and willingly shared their treasures of photographs, newspaper articles, family stories, souvenirs and letters. The History Book Committee spent many long hours in meetings in all kinds of weather to collect, evaluate and organize the materials, write and proof text while maintaining a sense of humor. Committee members were: LeDon Young and Norm Lathbury, Co-Chairs; Mel Brown, Darlene Confer, Maryann Haagen, Joe Hartle, Shirley Heidrich, Anna (Mickey) Peters and Lee Pressler.

The process of publishing a quality book is an arduous undertaking. It requires a production schedule with deadlines that must be met in order to have the final copy ready for the printer. To this end we owe Centre Publications and their staff a debt of gratitude. Gene Kneller spent many hours providing guidance and suggestions to the committee. His expertise was the glue that made this project come together.

Finding accurate documentation is the most difficult task in any historical writing. While it is the intent to portray events and the people involved accurately, problems arise when there is conflicting information. Therefore, the committee agreed that we should use the most contemporaneous sources available - they were newspaper articles, minutes, and other written documentation. We apologize for any errors or omissions.

Thanks also to Jim Biddle, Mark Lucas, Ray Sharer, John Wert, Jr. and Barbara Brueggebors, whose knowledge of the Fair's history helped guide the organization and development. We are indebted to the *Altoona Mirror*, the *Centre Daily Times*, the *Centre Democrat*, the Aaronsburg Historical Museum and Penn State's Pattee Library for access to the archives of the *Pennsylvania Mirror*, the *Centre Reporter* and the *Millheim Journal*. Thanks also to Jeff Wert, Ron Flood and Kathy Lathbury for their constructive criticism of the manuscript.

The committee owes a special debt of gratitude to the author, Norm Lathbury, who spent countless hours researching, writing the manuscript, working with the publisher to insure continuity and keeping the production schedule on track.

We hope you enjoy this tribute to the 125th Anniversary of the Nation's most unique agricultural county fair, the Grange Encampment and Centre County Fair.

The History Book Committee

"Until 1918 no admission was charged. But it was decided that it would be in the best interest of the camp to charge a small fee as more income was required from year to year."

Table of

GRANGE PARK

Contents

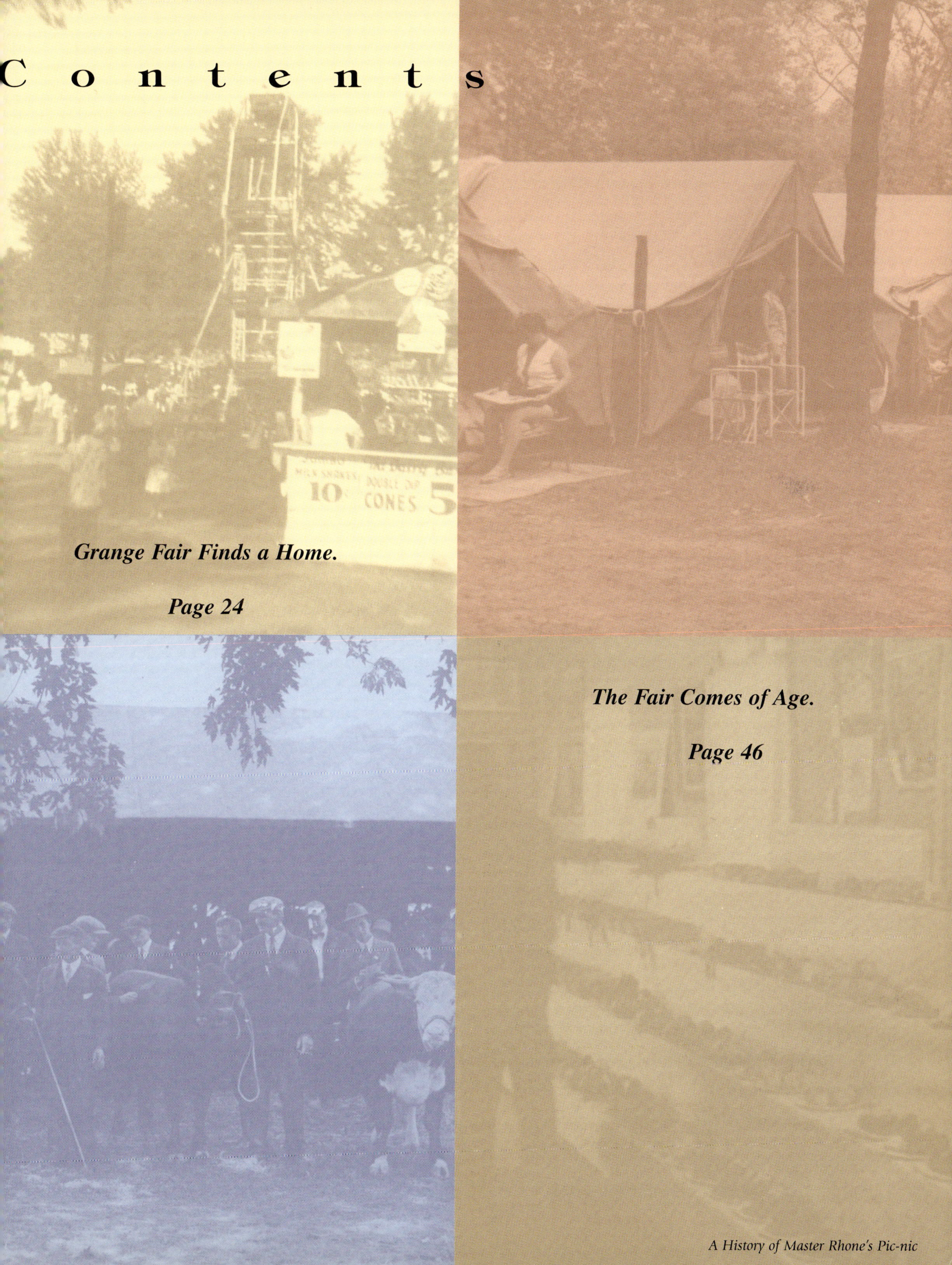

THE GRANGE: Why, When and How.

The old saying goes – "Despite all our accomplishments, humanity owes its existence to a six inch layer of topsoil and the fact that it rains."

It has always been the farmer, through long hours of lonely, arduous physical labor, at the mercy of Mother Nature and those to whom agricultural goods are sold, who have been the backbone of this nation. Following the Civil War and the Reconstruction period, mechanization, and scientific methods began to be applied to agriculture. The newly formed Grange played a significant role in educating and organizing farmers and their families.

To tell the story of how the Grange Encampment and Centre County Fair began, without explaining how the Patrons of Husbandry or Grange was established, would be ignoring a very crucial part of the story.

At the close of the Civil War in 1865, our nation embarked upon the overwhelming task of reconstruction: the rebuilding of the country in order to survive. Industries were abandoned or destroyed, without raw materials and manpower, as a result of the war. Agriculture was virtually at a standstill, especially in the South and West. The nation's food supply had to be promptly revived and production returned to pre-war levels. Soldiers and sailors from both sides returned to their hometowns to resume their lives, but life would never be the same. Many turned their energies to agriculture because it appeared to be a way to find some economic relief. Western farmers were banking on the promises of railroad owners. Many southern, northern and western farmers had little or no education and were often taken advantage of by middlemen and the railroad agents who bought the fruits of their labor at below market prices and promptly sold them for a handsome profit. No matter how hard families toiled and struggled, they saw no tomorrow, no future for their dreams. Yet another burden was isolation. In many instances, neighbors were few and far between. There was little opportunity to "talk things over." Working alone resulted in an atmosphere of hopelessness and despair.

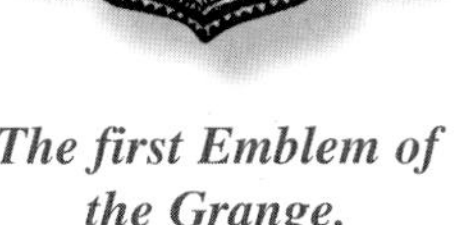

The first Emblem of the Grange.

Now enter into the picture, Oliver Hudson Kelley, a Massachusetts native who established a homestead on the headwaters of the Mississippi River near Itasca, Minnesota. Kelley supplemented his meager farm income by trading with the Indians and writing many articles on farming. His writing encouraged farmers to "add more study to their physical labors and experiment and record their findings." Articles that he wrote became so popular that he became a regular contributor to the Federal Department of Agriculture publications. He was offered and accepted a clerkship in the department in the winter of 1864.

Kelley returned to his Minnesota farm in the spring of 1865, but was summoned back to Washington by Agriculture Commissioner Isaac Newton.

Oliver Hudson Kelley, principle founder of the Patrons of Husbandry.

Kelley was informed that President Andrew Johnson wanted a survey made of agricultural conditions in the South, as a means of hastening the rehabilitation of the nation's basic industry and solving the grave reconstruction problem. Kelley was commissioned on January 1, 1866, and soon began his trip. What he discovered was appalling. The Southern farms were devastated, destroyed or in varying states of disrepair. Similar conditions existed in some areas of the North and the West.

How to solve this overwhelming situation on three fronts would require a significant effort. Kelley had a germ of an idea. He had toyed with it before. What would happen if there were nationwide fraternal organizations of farmers? After returning to Washington, he set out to make his dream come true. Kelley realized that he could not do this alone. He enlisted the help of six of his fellow associates in various government departments and on November 15, 1867, the name "Patrons of Husbandry" as well as the word "Grange" to designate local units was adopted. A meeting of the founders was held in Washington D.C. on December 4, 1867, and they declared themselves the National Grange of the Patrons of Husbandry.

The plan of the founders was simple. First, a local community association was established, consisting of neighbors and friends to enable discussion of local issues with candor and tolerance. Second, to provide a forum to understand more clearly community needs and mutual methods to resolve them. Third, transforming strangers into neighbors, enemies into friends, and in the process, clarify misunderstandings, remove prejudices, provide a basic level of community education and establish the spirit of cooperation. This was to become the Subordinate Grange, the local unit which would be the basis of the county, state and national organization. The Grange would ultimately become one of the most influential movements of the times leading the restoration of national unity, stability and good will.

While the early years were very difficult and disheartening, including the near collapse of the Order in the late 1870's, the Grange slowly began to grow. In Centre County, a local farmer and visionary, Leonard Rhone was instrumental in establishing the Grange. Leonard Rhone was born on July 21, 1838, son of Jacob and Sarah Rhone. He was raised on a farm 3 miles west of Centre Hall that was purchased by his grandfather, Michael Rhone in 1794. (The farm house and some of the out buildings still stand on the estate known today as "Rhoneymeade" on Rimmey Road.)

Rhone worked on his father's farm during the summers and attended public school in the winter. After his father died in 1853, he continued farming operations with his mother, but strongly believed that he needed to further his education. He enrolled at Kishacoquillas Academy where he dutifully studied agriculture. In 1864, Rhone married Maggie Sankey, daughter of James Sankey of Potters Mills. In 1865 his mother relinquished the farm and Rhone became its sole proprietor. During the decade that followed, Rhone recognized the need for a unifying organization among the farmers. Like Oliver Hudson Kelley, he, too, was keenly aware of the isolation, the need for education and socialization. Other professions and vocations realized the benefits of combined efforts and there wasn't any reason that farmers should not or could not take advantage of a fraternal brotherhood. When he learned of Kelley's efforts to organize the Patrons of Husbandry and the Grange, he took immediate action.

The Rhone Homestead

Rhone quickly recognized the benefits and the power of the Grange. He initiated and encouraged the organization of Progress Grange and other subordinate Granges in Centre County. By December 1874, sixteen Granges had been chartered.

Progress Grange #96 is shown above. The structure was built in 1898.

The Subordinate Granges of Centre County

Of the original number, nine remain active under their original charters. They are:

- Progress Grange # 96, Centre Hall, February 3, 1874
- Logan Grange # 109, Pleasant Gap, February 13, 1874
- Victor Grange # 325, Oak Hall, March 10, 1874
- Washington Grange # 157, Pine Hall, March 23, 1874
- Halfmoon Grange # 290, Stormstown, June 19, 1874
- Bald Eagle Grange # 151, Runville, June 30, 1874
- Howard Grange # 227, Howard, July 7, 1874
- Marion Grange # 223, Jacksonville, July 24, 1874
- Union Grange # 325, Unionville, August 11, 1874

Other Granges that were organized in 1874 but are now dormant:

- Benner Grange # 207
- Providence Grange # 217
- Spring Mills Grange # 158
- Centre Grange # 254
- Excelsior Grange # 283
- Fairview Grange # 296
- Walker Grange # 245

Granges that were organized from 1881 to 1905 *except Port Matilda* and are now dormant:

- Zion Grange # 757
- Oak Grove # 761
- Leonard # 779
- Goodwill # 1050
- Miles # 1051
- Madison # 1053
- Romola # 1192
- Moshannon # 1272
- Port Matilda # 1284

Granges organized from 1916 to 1936 were:

- Penn State # 1707
- Rebersburg # 1919
- Pine Glen # 1981
- Baileyville # 1991
- East Penns Valley # 2000
- Walker # 2007

Walker and Baileyville still operate under their original charters. Spring Mills and East Penns Valley combined and was renamed Penns Valley Grange. The remaining Granges listed above are now dormant.

The Minutes from Charter Meeting of Progress Grange #96 Establishing the First Subordinate Grange in Centre County.

Centre Hall Feb 3d 1874—
The following persons met in the vacant School room in Centre Hall— J. J. Arney— James A Keller. Ephram Keller. David Rhinesmith, G. M. Boal. Leonard Rhone. Lafayette Neff. James A Lingle Samuel Crotzer. John. B. Bittner. Daniel Irvin N. D. Osmen Geo. Hoffer. Thomas Lingle ~~A H Hosterman~~ James H. McCormic. Petre Smith Daniel Fleisher.
Maggie Rhone Maggie E Keller Mary A Neff— Sallie J Arney. Barbry Bittner Sarah Sankey. Susan M. Hoffer. Maggie E Hoffer. Mary J Rhinesmith (The following Charter members where absent— W. A. Boal A. H. Hosterman and Lottie K Keller.—)
And were organized into a Subordinate Grange of the Patrons of Husbandry— by Deputy Frank Porter of Eagle Grange No1, Assisted by W. R. Bierly of Williamsport Pa.
The following persons were elected as officers for the ensuing Year Master J. J. Arney Overseer Leonard Rhone, Lecturer Dr. Petre Smith Steward John Sankey—, Assistant Steward David Rhinesmith Chaplain Daniel Fleisher: Treasurer. G. M. Boal Secretary James A Keller. Gate Keeper. Geo Hoffer, Ceres Maggie Rhone Pomona Mary A ~~Neff~~ Flora Maggie E. Keller; Lady Assistant Steward Maggie Hoffer. Trustees James H. McCormic: Petre Smith and James A Lingle— On motion the name Progress was adopted for the Grange. The Grange adjourned to meet in the Evening at 7 Oclock P.M
J. A. Keller Sec.

Centre County Pomona Grange was chartered on September 15, 1875, which included all of the Subordinate Granges in the county. Rhone was elected Pomona Master and was re-elected for five consecutive terms, a period of twelve years. Other officers elected were: M. P. Weaver, Overseer; John Rishel, Lecturer; J.J. Musser, Steward; J.H. Keller, Assistant Steward; H.L. Harvey, Chaplain; John H. Grove, Treasurer; James P. Weaver, Secretary; George Hoffer, Gatekeeper; Susan Musser, Pomona; Rose Rishel, Ceres; Sara Taylor, Flora; Lady Assistant Steward, Kate M. Dale; Executive Committee: George Dale, H.L. Harvey, J.J. Arney, J.F. Weaver, and George Campbell; Finance Committee: W.F. Rearick, George Taylor and Jacob Dunkle.

Rhone's undaunted dedication and enthusiasm won him an appointment as State Deputy in 1875, a position he held for seven consecutive years. Rhone also served as Master of Progress Grange from 1877-1879, and was elected Overseer of the Pennsylvania State Grange in 1878. At the close of his term as Overseer in 1880, he was unanimously elected as State Master at the State Session held in Greensburg, Westmoreland County. Rhone reigned as State Master for eighteen consecutive years, until 1898, longer than any other State Master to date.

Rhone's accomplishments were many. Perhaps more than anything else, none was as significant as his commitment to the socialization and education of the agricultural community. Rhone was particularly concerned with the impoverished living conditions, the loneliness of rural life and the need for farmers to form cooperatives in order to prosper. Cooperatives provided both buying and selling power enabling farmers to sell their crops at a higher prices and purchase supplies and equipment at lower prices by eliminating the middleman. Rhone was also instrumental in bringing rural mail service or what was know then as Rural Free Delivery (RFD) to the Centre Hall area. Rhone's biography begins on the next page as it appeared in *Linn's History of Centre County*.

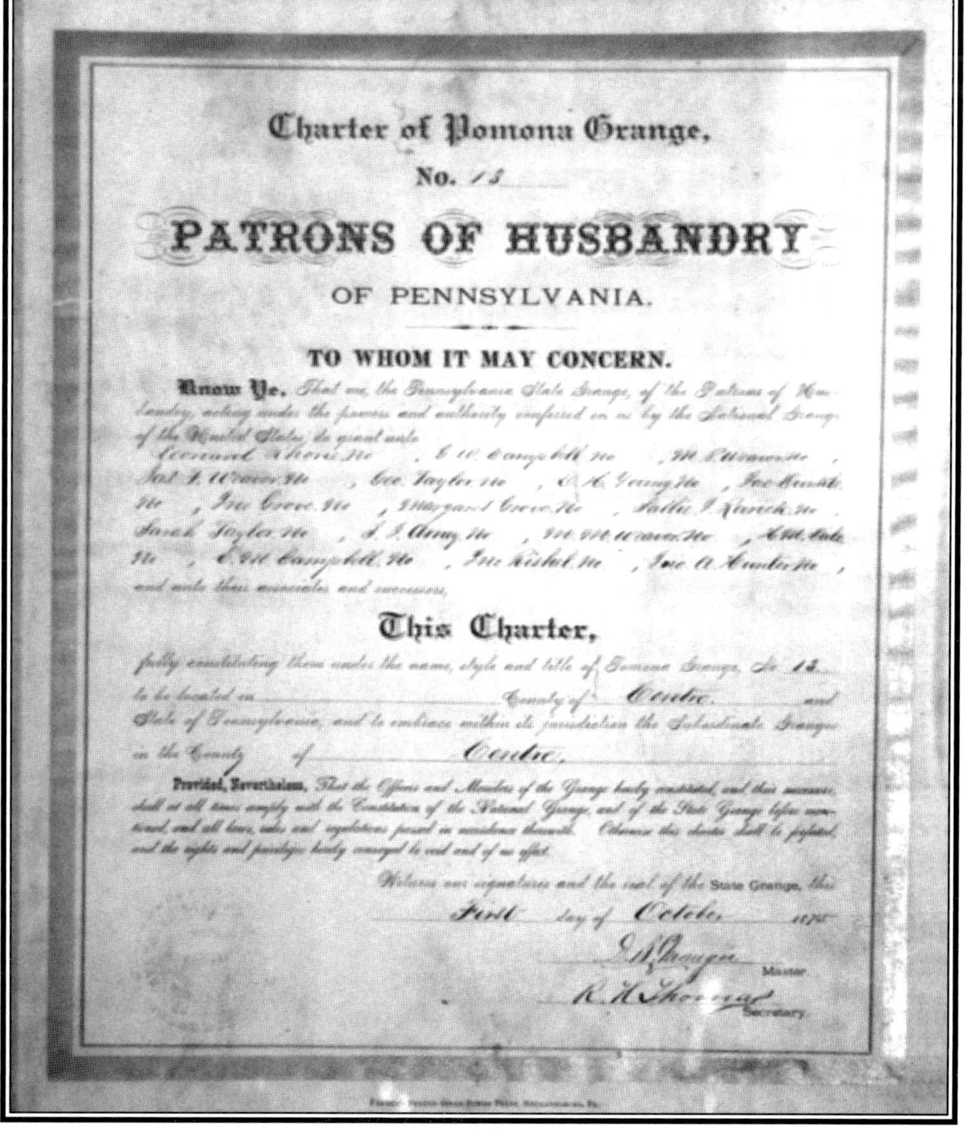

Charter of Pomona Grange,

No. 13

PATRONS OF HUSBANDRY

OF PENNSYLVANIA.

TO WHOM IT MAY CONCERN.

Know Ye, ...

This Charter,

... County of Centre ...

... in the County of Centre.

Provided, Nevertheless, ...

... State Grange ...

First day of October 1875

Master

Secretary

LEONARD RHONE, ESQ.

From Linn's History of Centre County

Leonard Rhone, Esq., was born on the farm on which he now resides, part of the old manor of Nottingham, once owned by the Penns, on the 21st of July, 1838. The farm is a part of No.1 of the divisions of the manor sold by the Penns to Jacob Straub by deed of June 24, 1794. Michael Rhone, Leonard's grandfather, purchased it Sept. 15, 1794, removing thither from the eastern end of Penn's Valley, and it has been in possession of the family ever since.

Leonard was one of the sons of Jacob and Sarah Rhone, and during his early life worked upon his father's place during the summer and attended a public school in the winter, thus acquiring a taste for farm-life, to which he is devotedly attached, and at the same time obtaining the rudiments of an education which he has not failed to vastly improve.

His father died in 1853, and Leonard remained on the farm with his mother, who with true womanly courage continued to carry on its operations, keeping the family together, and securing for them all the educational advantages the neighborhood afforded until its members arrived at a proper age to enter upon higher courses of study at seminaries and colleges.

In 1857, Mr. Rhone served a voluntary apprenticeship of some months in a coach-maker's shop, in order to acquire a knowledge of the use of tools, with a view to enable him to make ordinary farm repairs. He readily acquired a knowledge competent for all such purposes. Deeming a better education in the sciences and learning of the schools of great importance, in November, 1857, he entered Kishacoquillas Academy as a student. When asked by Mr. Alexander, the principal, what pursuit or profession he proposed to select for life, he promptly responded, "that of a farmer."

After pursuing his studies (in which he made great progress) for a year, he was again approached with the question of his intended pursuit, with the suggestion that greater opportunities for distinction

Leonard Rhone

1794. MICHAEL RHONE. 1846. JACOB RHONE. 1894. LEONARD RHONE.

Centre Hall, Pa.

The honor of your presence is requested at the Rhone Homestead.

On Wednesday Morning, June 13, 1894, Ten o'clock.

Celebration of the One Hundredth Anniversary of the occupancy of the Homestead by the Rhone Family.

MRS. SARAH RHONE, (Mother).

J. W. RHONE. LEONARD RHONE. JENNIE RHONE MASON.
MARY RHONE DALE. JACOB RHONE. ALICE RHONE HIXON.
SARAH RHONE HESS. EMMA RHONE SANKEY.

The celebration of the One Hundredth Anniversary of the Rhone Homestead illustrates Master Rhone's commitment to the stewardship of the land and to his neighbors in communities near and far.

One Hundredth Anniversary
OF THE OCCUPANCY OF THE
RHONE HOMESTEAD

Formal Reception, 10 A. M.
Music.
Address of Welcome, Leonard Rhone.
Response, Mr. H. H. Harshberger
Music.
Historical Address, Col. Jas. F. Weaver.
Music.
Anniversary Recitation, Miss Emma Brewer.
Music.
Dinner, 12 M.

RAILROAD STATION GREGG.

TRAINS
Arrive from the West at.......... 7:00 A. M.
Arrive from the East at.......... 8:16 A. M.
Leave for the West at.......... 4:13 P. M.
Leave for the East at.......... 2:54 P. M.

CARRIAGES WILL MEET ALL TRAINS.

awaited professional men. But firm to his resolution, he refused to give up his choice of a calling. His bent was shown in his essays, which were upon agricultural subjects, and his exhibition oration had for its topic "Agriculture."

During the winter of 1858-59 he taught school, thus firmly grounding his knowledge in efforts to instruct others, and at the close of his school returned to his home and assisted his mother on the farm.

In 1864, Mr. Rhone married Miss Maggie Sankey, daughter of James Sankey, Esq., of Potter's Mills, a lady of extraordinary energy of character and rare mental worth. His mother relinquishing charge of the farm, he became lessee of the old homestead, and in May, 1865, upon a sale of the place on mutual agreement of the family, he became sole proprietor of his father's property.

During these years of farm-life, while actively engaged in his favorite pursuit, he began to feel the importance and necessity of some organization among farmers of a social and educational character, for the purpose of breaking up the monotony of rural life and cementing the tillers of the soil together in one fraternal brotherhood. This was not only a theory with him, but a real principle. He felt it, and as he pondered over the situation the idea burned itself into his very soul, and he longed for the day when the isolation of the farmer and his family should be broken up, and they could meet upon a fraternal platform for mutual benefit and protection. He saw other pursuits and professions banding together for like purposes, realized the benefits of combined efforts to their respective classes, and earnestly desired to see the same principles introduced among the farmers. Whilst absorbed with this thought the Patrons of Husbandry began to formulate their ideas in the same direction. As their work unfolded he discerned the wisdom of the new organization, and he watched its rise and progress with the most intense interest. As it expanded, and like a tidal wave rolled on, he saw its adaptation to the situation, and determined to enter into the work. In connection with a few of his neighbors, application was made for a charter, and on the 3d day of February, 1874, Progress Grange No. 96, was duly organized at Centre Hall, he being a charter member of the first subordinate grange instituted in Centre County.

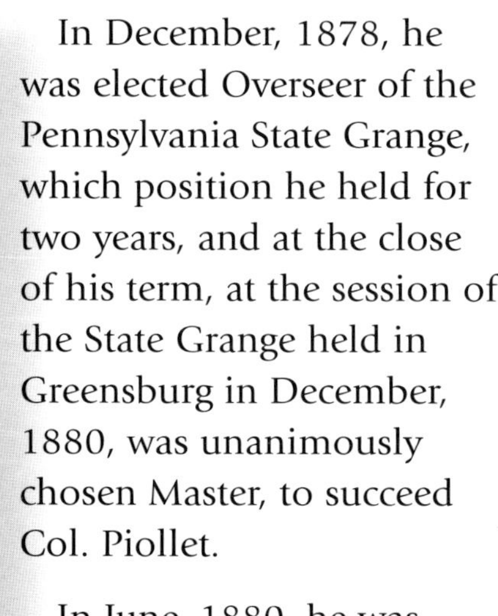

Postcard, circa 1915.

Once inside the gate, he saw still more clearly the power of the organization for good. He applied himself with great zeal and energy in the new field, so faithful and earnest were his efforts that on the 9th of April following he was commissioned by D. B. Waugen, Master of the Pennsylvania State Grange, deputy for Centre County, which position he continued to hold by consecutive annual appointment for seven years, during which time he displayed most untiring activity and performed an immense amount of labor in behalf of the organization. And so successful were his labors that the organization in Centre County today is admitted to be more complete than in any other county in the State.

At the organization of Centre County Pomona Grange, No. 13, Sept. 15, 1875, he was elected Master for one year, and re-elected five consecutive terms. In December, 1877, he was chosen Master of his own grange, No. 96, and served his term with great acceptability.

In December, 1878, he was elected Overseer of the Pennsylvania State Grange, which position he held for two years, and at the close of his term, at the session of the State Grange held in Greensburg in December, 1880, was unanimously chosen Master, to succeed Col. Piollet.

In June, 1880, he was elected a trustee of the Pennsylvania State College, located in Centre County, which position he has filled with marked ability and faithfulness, and impressed his ideas to some extent upon the workings of the institution.

In September, 1880, the National Greenback-Labor party nominated him for the State Legislature, but having no proclivities for political life he declined on account of private business and other pressing engagements in connection with his position as an officer of the State Grange.

Some time prior to the assembling of the National Agricultural Convention, which met in the city of New York in January, 1882, Governor Hoyt, unsolicited, appointed him a delegate to that body. Previous engagements, however, prevented him from attending.

In stature Mr. Rhone is five feet eight inches high, erect, compactly built, and capable of great endurance. The immense amount of labor he performs as Master of the State Grange, in correspondence, public addresses, etc., in connection with his own private business, is sufficient to break down an ordinary man, but he bears it well, and seems to improve under the severe tension to which his physical ability is subjected.

Intellectually, Mr. Rhone is a plain, practical thinker. He only arrives at conclusions after mature consideration. He may justly be considered a safe counselor. His administration of the State Grange has been eminently successful. As a presiding officer he is calm, courteous, and firm, and wields the gavel with grace. As a speaker he is diffident, but only appears before an audience after mature preparation. Some idea of his style may be obtained by a short quotation from one of his addresses:

"Agricultural success is pre-eminently essential to national prosperity. It affords employment to over six hundred millions of the human race, and subsistence to all the multiplied millions that dwell upon the earth. Should Heaven for a single season frown upon the efforts of the farmer, by withholding its rain and sunshine, or by sending mildew and blight, the general round of prosperity in every department of government and trade would be silent as death; machinery would become motionless; vessels would be tied in their moorings; the efforts of genius would be paralyzed; the distortions of want and despair would fill the places of departed plenty and gladness. Yet, notwithstanding all these facts, the importance of agriculture is not properly appreciated. Our government is slow to recognize its real magnitude and worth. Appropriations of money for its promotion and development are small and insignificant compared with the fostering care and protection afforded to other and minor interests."

Personally, Leonard Rhone is a warmhearted, generous, and true man. Never forgets a favor or forsakes a friend. Appeals to his generosity are never made in vain. His moral character is unsullied and his Christian conduct above reproach. In a word, he is one of Nature's noblemen.

On February 3, 1874, in an empty room in the Centre Hall School, Centre County's First Grange, Progress Grange # 96 was chartered. The school was located on the property of a residence, which is now surrounded by the Lutheran & Reformed Cemetery at 325 Church Street.

THE PIC-NIK: A Tradition Begins.

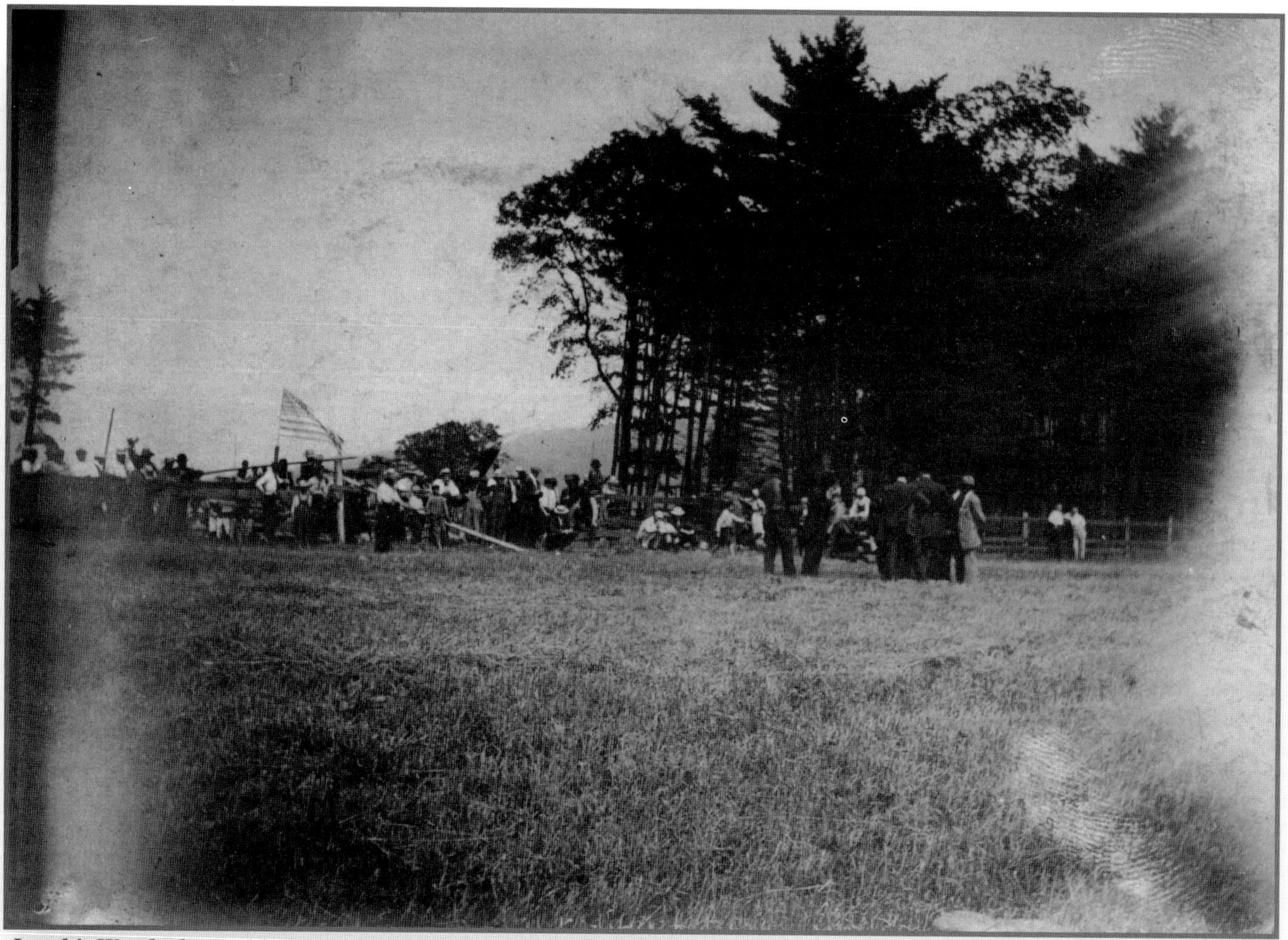

Leech's Woods, located between Centre Hall and Linden Hall, was the site of the first pic-nik. More than 3,000 folks journied to the "Old Camp Meeting Ground" on September 24, 1874. The gentleman in the gray suit in the foreground at the right of the photo is Master Rhone.

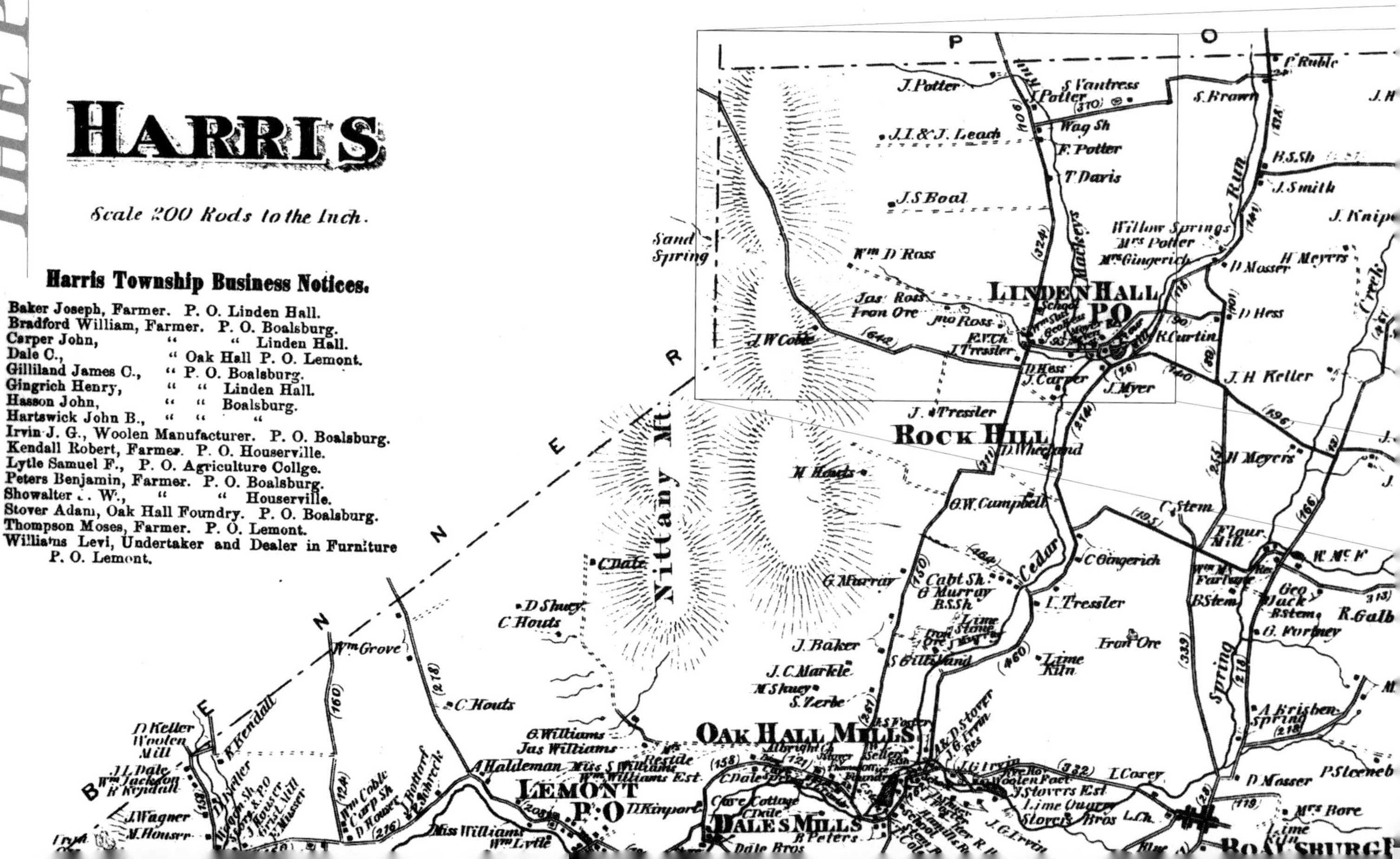
HARRIS
Scale 200 Rods to the Inch.
Harris Township Business Notices.
Baker Joseph, Farmer. P. O. Linden Hall.
Bradford William, Farmer. P. O. Boalsburg.
Carper John, " " Linden Hall.
Dale C., " Oak Hall P. O. Lemont.
Gilliland James C., " P. O. Boalsburg.
Gingrich Henry, " " Linden Hall.
Hasson John, " " Boalsburg.
Hartswick John B., " " "
Irvin J. G., Woolen Manufacturer. P. O. Boalsburg.
Kendall Robert, Farmer. P. O. Houserville.
Lytle Samuel F., P. O. Agriculture Collge.
Peters Benjamin, Farmer. P. O. Boalsburg.
Showalter . . W., " " Houserville.
Stover Adam, Oak Hall Foundry. P. O. Boalsburg.
Thompson Moses, Farmer. P. O. Lemont.
Williams Levi, Undertaker and Dealer in Furniture
P. O. Lemont.
LINDEN HALL P.O.
ROCK HILL
OAK HALL MILLS
LEMONT P.O.
DALES MILLS
Nittany Mt
J. Potter
S. Vantress
J. I. & J. Leach
Wag Sh
F. Potter
T. Davis
J. S Boal
Willow Springs
Mrs Potter
Mrs Gingerich
Wm D Ross
Jas. Ross
Iron Ore
Jno Ross
J. W Cobb
J. Tressler
J. Carper
J. Myer
D. Whitehand
R Curtin
D Hess
D Mosser
H Meyers
J. H Keller
S. Brown
B.S.Sh
J. Smith
Sand Spring
G.W. Campbell
C. Stem
Flour Mill
C Gingerich
Cedar
I. Tressler
G. Murray
Iron Ore
Lime Kiln
J. Baker
J. C Markle
S. Zerbe
C. Dale
D Shuey
C. Houts
Wm Grove
C Houts
G. Williams
Jas Williams
A Haldeman
Miss Williams
D Keller
Woolen Mill
J. Wagner
Lime Quarry
Stovers Bros
I. Corey
D Mosser
P Steeneb
Mrs Hare
A. Krisben
Spring
R. Galb
Geo Jack
G. Fortney
Dale Bros
B. Peters
J. G. Irvin

So, where did the idea for the Grange Fair come from? Was it always a fair or was it something else? Those of you who have been attending this event for fifty, sixty, seventy years or more know that it was, and still is referred to as "Grangers Picnic" or just "Picnic." The idea or notion to have a picnic or as it was spelled in 1874, "pic-nik," came from Leonard Rhone. According to the Minutes from Progress Grange on August 23, 1874:

"On motion of Brother Rhone we join with Sister Granges in holding a pic-nik was agreed to and the following Com. Appointed to make the necessary arrangements, fix the Grounds etc., Brothers Rhinesmith, Neff and G.M. Boal. After consultation, it was agreed that the pic-nik be held on the Old Camp Meeting Grounds near Joshua Potters on the 24th day of September." The Potter Farm was located on the Linden Hall Road about 3.5 miles West of Centre Hall.

Centre Reporter September 24, 1874 (Thursday) -- First Centre County Grangers' Pic-nic.

Held in Leech's Woods, near Centre Hall, it was a great success. Enjoyed by all of the nearly 3000 who attended. At 11:30 a. m. following officers were elected:
President: Christian Dale, Sr. of Victor Grange.
Vice-Presidents: John Rishel, Progress Grange; G. W. Campbell, Victor Grange; A. J. Thompson, Halfmoon Grange; M. P. Weaver, Logan Grange; R. G. Brett, Centre Grange; George Taylor, Bald Eagle Grange; Christian Alexander, Providence Grange; Jas. M'Clintick, Spring Mills Grange; Robt. M'Night, Benner Grange; Henry Sankey, Fairview Grange; John Musser, Potter Grange and Jacob C. Smith.
Secretaries: Christian Dale, Jr., Union Grange and David Young.

Exercises were then opened by prayer by Rev. Hartsock of Boalsburg. Deputy Leonard Rhone delivered the address of welcome after which Capt. John A. Hunter, of Halfmoon Grange, was then introduced as the speaker. He spoke of the origin of the Patrons of Husbandry and their objectives, the necessity for farmers to pay closer attention to the advancement of agriculture and the importance of making their farms less irksome to their sons and make agriculture more attractive to them.

At noon eight or ten acres of the woodland were dotted with many hundreds of tablecloths and a feast was enjoyed by all. Following dinner a parade formed, headed by the Boalsburg band and the Pine Grove Mills Band in the centre. The Marshalls, heading the procession, were Col. Weaver and Capt. Geo. M. Boal.

Following the parade the Declaration of Principles was read by Henry Keller followed by address by Col. James F. Weaver of Milesburg.

The event was the largest of the kind ever held in the county and will be long remembered.

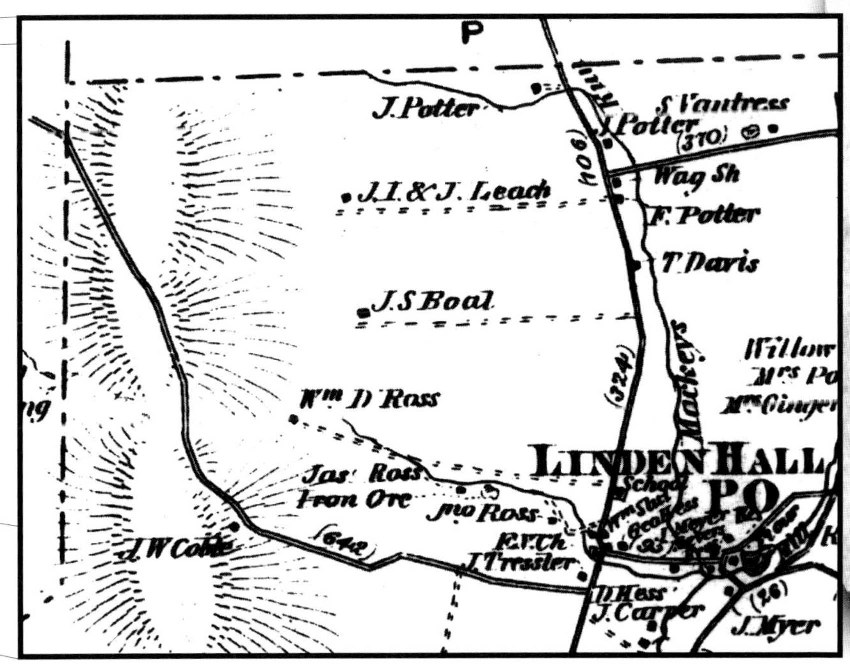

Taken from Pomeroy's 1874 Atlas, the maps shown illustrate the location of the first "Pic-nik." Note that the spelling of LEECH is incorrect on the map, a common error in that time.

Traveling by buggy on Centre Hall Mountain, circa 1920.

The Reporter August 24, 1876
The next annual Granger's pic-nic of Centre County, will be held on Nittany mountain, near Centre Hall, on Sept. 28. It will no doubt be the largest affair of the kind yet held by the patrons.

September 21, 1876
The Grangers of Centre County will hold their third annual basket pic-nic on Thursday, Sept. 28th on top of Nittany mountain on the turnpike 1 mile west of Centre Hall. D.R. Mauger, master of Penn'a State Grange and other prominent speakers of the order have been invited and are expected to be present. Centre Hall and Farmers Mills Cornet Bands have been engaged for the occasion. The day is to be one of recreation and pleasure and free to all; a general invitation is extended to all who may have any desire to be present.

From 1876 to 1887, the Pic-nic was held on the top of Mt. Nittany, Centre Hall Mountain, with the exception of 1878. Below, the Snyder's pose with their picnic fare.

About 1885 several families brought tents to enable them to attend the day-long festivities and travel safely home in daylight the following day. In 1887 Master Rhone borrowed old Civil War tents from the Pennsylvania National Guard in order to accommodate more families. This was the beginning of tenting, exhibits, and a three day fair. In 1888, the dates were changed from September to August. Centre County's Grange Fair is now the only encampment and fair in the nation.

October 3, 1876

The grangers of Centre County held their third annual pic-nic, on Nittany mountain, near this place, on Thursday last, 28. The weather was all that could be wished, and the spot selected one of the most delightful in the state. Early in the morning strings of wagons and buggies came pouring through Centre Hall from the valley, loaded with men, women and children, and baskets of the best eatables the valley can afford. The turn-out from the other side of the county was about as large as from over here, and at noon there were near 3,000 persons on the ground.

The exercises in the morning were opened by prayer by H. L. Harvey; then followed an address of Welcome, by deputy Leonard Rhone, after which the Declaration of Purposes of the order was read by Col. Jas. F. Weaver, when the Chairman, Mr. G. W. Campbell, announced an intermission for dinner, when it was expected the speakers invited from a distance would arrive. The crowd then dispersed into a hundred groups and dotted the woods with cloths spread upon the grounds and the feast commenced. There was an abundance of good things and baskets full left. The assembly having been called to order again, after dinner, it was announced that the expected orators had failed to make their appearance, and that the home talent of the Order would have to be fallen back upon to deliver addresses. The first speaker called upon the stand was Col. Jas. F. Weaver, who spoke about 3/4 of an hour upon the objects of the Order. The speakers who followed were deputy Rhone and H. L. Harvey, in appropriate but brief addresses when the assembly was dismissed, all having enjoyed a pleasant time. Excellent music was furnished during the day by the Farmer's Mills and Centre Hall brass bands.

Fair Dates & Locations

According to newspaper accounts and minute books the following list illustrates the location and the date(s) from 1874 to 1889.

September 24, 1874	- Old Camp Meeting Grounds at Leech's Woods -near Potter Farm
September, 1875	- Centre County Fairgrounds, Bellefonte
September 28, 1876	- Mount Nittany, Centre Hall Mountain
September, 1877	- Mount Nittany, Centre Hall Mountain
September 19, 1878	- Pennsylvania State College
September 25, 1879	- Mount Nittany, Centre Hall Mountain
September 16, 1880	- Mount Nittany, Centre Hall Mountain
September 15, 1881	- Mount Nittany, Centre Hall Mountain
September 21, 1882	- Mount Nittany, Centre Hall Mountain
September 20, 1883	- Mount Nittany, Centre Hall Mountain
September, 1884	- Mount Nittany, Centre Hall Mountain
September, 1885	- Mount Nittany, Centre Hall Mountain
September, 1886	- Mount Nittany, Centre Hall Mountain
September, 1887	- Mount Nittany, Centre Hall Mountain
August, 1888	- Old Fort Woods, Centre Hall
August, 1889	- Old Fort Woods, Centre Hall
1890 to 1999	- Grange Park

The fair was held to a one day gathering in 1943 due to gasoline rationing during World War II.

The first accounts of the "Pic-nik" are from The Centre Reporter, the local newspaper. Some controversy exists as to when the first "Pic-nik" was held. According to some old newspaper accounts the first "Pic-nik" was held in the summer of 1873. Further research reveals that this was a picnic of the Centre County Farmers Organization, which was not connected with the Grange.

There have been many lively discussions about the locations of the early "Pic-niks." The table on the opposite page was prepared using the resources of newspaper accounts and minute books.

Exhibits and organized tenting first appeared in 1887. Tents were borrowed from the Pennsylvania National Guard. Exhibits consisted of various pieces of farm machinery.

Above: Frances Harpster Homan (farthest right-standing) enjoys a day at the Fair in her pretty lavender lawn dress.

This family is preparing to have a meal at the 1910 Fair. The cookstove would have been shared with the families in the surrounding tents.

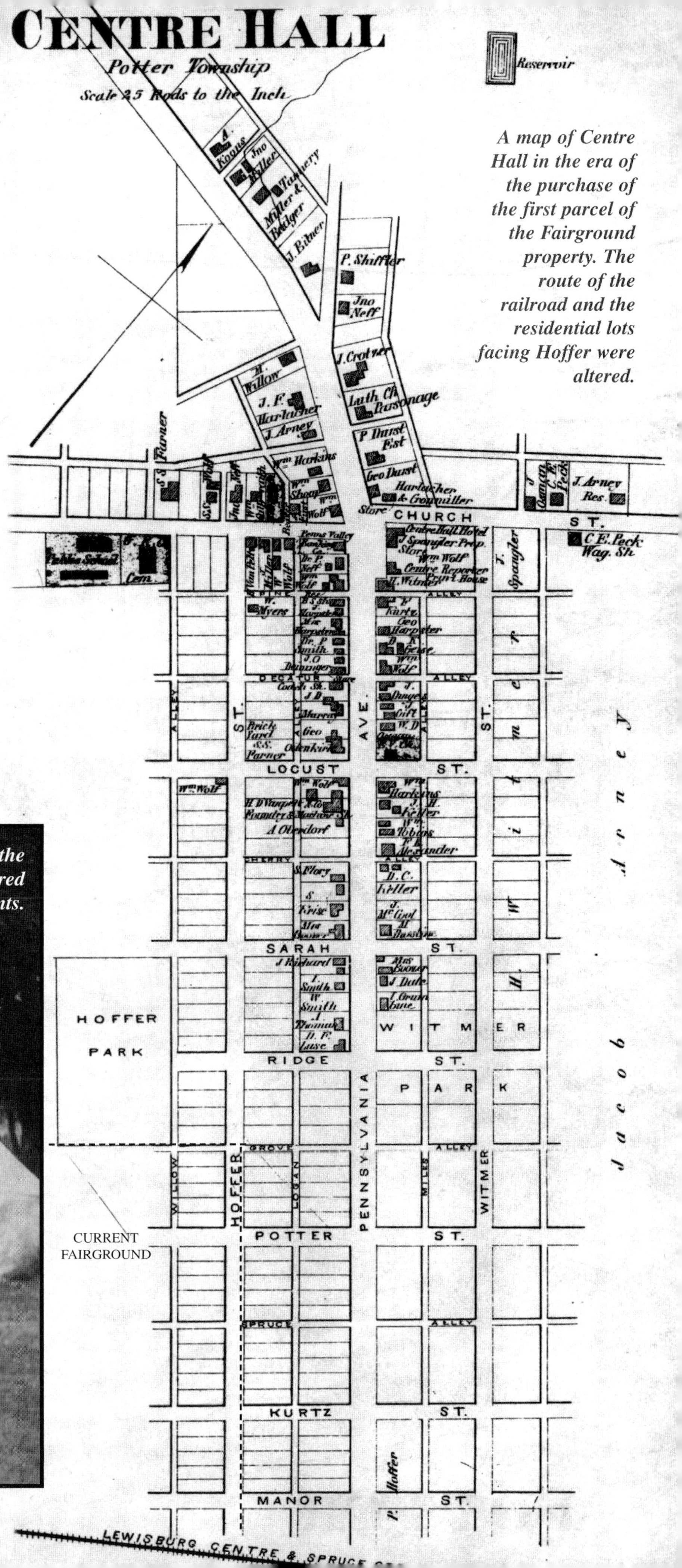

CENTRE HALL

Potter Township

Scale 25 Rods to the Inch

A map of Centre Hall in the era of the purchase of the first parcel of the Fairground property. The route of the railroad and the residential lots facing Hoffer were altered.

HOFFER PARK

CURRENT FAIRGROUND

LEWISBURG CENTRE & SPRUCE

Families getting together at the Fair is our oldest tradition. This 1910 reunion includes at least three generations.

In 1890 the "Pic-nik" was held at the newly acquired Grange Park located on the West Side of Centre Hall. The new site contained 26.25 acres and consisted of three separate parcels deeded to Centre County Pomona Grange Encampment and Fair. Twenty acres were purchased from John and Sarah Arney for $2,300 and six and one quarter acres from A.G. and Catherine Curtin for $739. The southern boundary was along the Lewisburg, Centre and Spruce Creek Railroad. The eastern boundary was along Hoffer Street and the northern and western boundaries along the properties of J.J. Arney and Jacob and Jennie Sharer. (See map.)

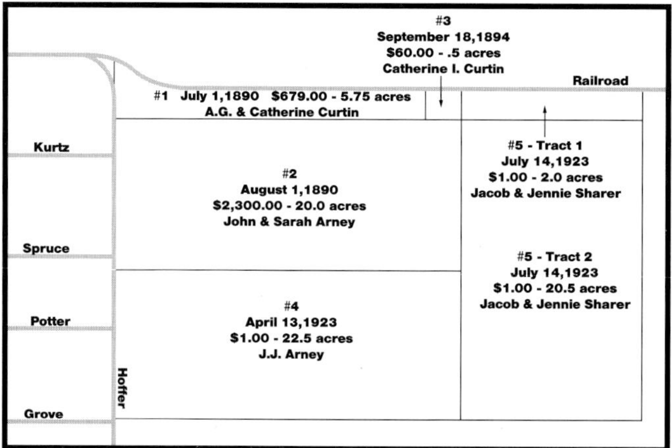

Plot Map of fair land purchases through 1923.

And what a year it was. The newly acquired fairground was nothing but wheat stubble. The weather was uncooperative with soaking rains. Tent stakes failed and tents collapsed in the mud. This, however, did little to dampen the spirit of the campers and did not keep the crowds from what was by now a well-established annual tradition. One of the early Grange historians estimated that five acres of ground was under roof and canvas that year. That was one-fifth of the total acreage.

Now that the "Pic-nik" had a new home, the task of organizing the park had to be undertaken. There are no written records or drawings to illustrate the original layout. Those responsible must have had some military background and training. The Headquarters Building was placed near the center of the park. The original building still stands, although it's not in the center of the park any more. It is, however, pretty much in the center of the tenting area. Tents were laid out in neatly organized rows which were given street names much like a military camp.

In 1914, all the county granges and other rural groups appointed permanent committee members for the fair.
In 1920, the custom of having department chairmen was instituted.

The Fair featured a display of old farm and household implements, as a reminder of the progress made by mechanization and education, as endorsed by the Grange.

In 1891 a permanent committee was selected to oversee the fair. Committee members were Leonard Rhone, chairman; George Dale, John Arney, George Boal, John Dauberman and George Gingerich. It is assumed that this committee took the responsibility for planning the layout of the fairground. In addition, they were instrumental in deciding what improvements needed to be accomplished. This would have included such things as road construction, tent locations, tree plantings and building and fence construction.

In 1914 Pomona Grange appointed permanent committee members. In 1915 the fair was expanded into a county fair and was incorporated as the Grange Encampment and Centre County Fair administered by Centre County Pomona Grange. Pomona Grange also elected candidates from Subordinate Granges to serve as Fair Board members.

The 1914 Fair Committee. From Left: Ralph Banney, Dan Grove, George Ralston, John Dale, Margaret Sankey, Jacob Sharer, Albert Spayd, David Keller and Clyde Dutrow.

Fair patrons and campers were mostly farmers from the Centre County area. Many of the activities consisted of lectures and demonstrations about farming and farm management. As Fair attendance continued to increase it became necessary to construct permanent buildings. One such building was constructed for meetings. This was the Auditorium, which is known now as the Commercial Building. Stables and other buildings were also erected. What started out as twenty-eight tents in 1877 was 200 tents by 1910. More exhibitors and concessions were part of the scenery.

Livestock, machinery, fruit, vegtables and other farm products were originally exhibited in tents.

Getting ready for the Fair took on a new dimension. Not only did tents have to be erected; each tent was equipped with a wooden floor, a table, a bench, and a bunk. You could even fill your own mattress with straw provided by the Fair if you brought your own tick. Cook stoves and firewood were also provided and shared by several tents. Most

Centre Reporter - September 1890
A cyclone struck September 16 at 4:00 pm. Much of the camp was blown away. Rain, severe lightning and thunder and steady rainfall lasted to 7:00 am. Sunny weather then took over and the camp was back in operation.

The agricultural exhibit of the Pennsylvania State College was housed in one of the first exhibit buildings constructed on the grounds. It is still in use today.

campers brought enough food for the week but storage was a problem and iceboxes weren't particularly practical. Several grocery stores on the fairground provided a solution to this problem.

A postcard of the bustling midway, circa 1905.

Electricity came to the Fair in 1916. Vernon G. Wagner and Elmer Stump installed the first electric lines into the park. A twelve-kilowatt transformer powered a few seventy-five watt bulbs to light the midway and main street. Tents were lighted with kerosene lamps and lanterns. As electric service improved, campers were able to rent a single light bulb for use in their tent for one dollar. For many campers this was their first experience with anything powered by electricity. It is interesting to note that only a select few Centre Hall residents had electric service. The surrounding rural areas were without electricity until the 1930's when the Rural Electrification Act took effect.

Sad news spread through the county as residents learned of Leonard Rhone's death. He died on September 17, 1917.

Attending the Fair was so important that even those traveling in Europe took extraordinary measures to get home in time for the Fair. Such an account is related in a letter written to John Knarr of Centre Hall by George W. Thomas. His letter appears on page 32.

Young ladies, dressed in "Gibson Girl" style, happily posed for this photo in 1907.

Some fair goers still can remember moving pictures being shown near headquarters. The poles shown in the foreground were used to suspend the screen for the evening showings.

August 15, 1929 - Millheim Journal
Week after next all roads will lead to Centre Hall where the Grangers will put on their annual encampment and fair. This fair grows in size and popularity and there is many a man who expect to "knock off work" for a week to camp on the fairgrounds. A week well spent in recreation, education and renewing acquaintances.

Grange Park was illuminated by kerosene lanterns which gave way to acetylene gas lamps starting in 1900.

Then as now, generations of families gathered at the fair.

This early postcard shows the newly built Auditorium on the far right. As income permitted more permanent buildings were erected and improvements were made.

Early concessions were "stick stands" covered by cloth or canvas. The ever popular Ferris wheel dominated the midway.

In 1912 Francis Harpster Homan (far right) traveled in a spring wagon from Rock Springs to the fair, a distance of some 25 miles.

Postcard, circa 1914.

Below is a letter written by George Thomas to John Knarr of Centre Hall August 24, 1914. Mr. Thomas was traveling in Europe at the beginning of the First World War.

Dear John:—

This epistle from war ridden Europe leaves us all well and none the worse off so far except the loss of one trunk which is in Germany and may turn up after the trouble is over.

At this writing I expect to leave here the 28th and unless caught by a German cruiser be with you all at Grangers.

There may be some mail forwarded to you from London for me. If there is kindly keep it for me till I put in an appearance.

Our trip this summer until the war chased us into London was a great success and both my grandmother and myself enjoyed Spain immensely. I took quite an interest in the bull fights while I was there. Talk about excitement. Well, you can find all you want at a Spanish bull fight.

Have been having a mighty good time in London regardless of the fact that the Germans are only sixty miles away. But having friends here makes a difference and there are a number of good things on at the theatres.

I suppose you know I'll land in Quebec and come to Centre Hall for the picnic via Montreal, Ottowa, Toronto and Buffalo.

Remember me to Jane and have the Grangers prepare plenty of fun for picnic week.

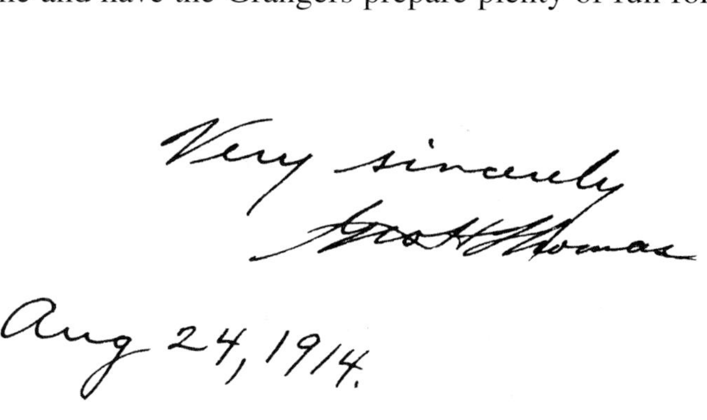

D.C. Keller's Boarding House provided home-cooked meals for hungry fair goers after a long day of taking in exhibits, livestock judging and visiting.

Even with the uncertain economy, as a result of the aftermath of World War I, the Fair continued to grow slowly through the 1920's. During 1923 the Fair Board purchased 45 acres consisting of two 22.5 acre plots. This increased the size of the fairgrounds to 70 acres. A network of electrical wires for the tent area was completed in 1927. Tenters could rent a light bulb for the duration of the fair for the tidy sum of $1.75. Two bulbs were $3.00. There was a deposit of 25 cents for each light bulb. The Fair also provided other items for rent as illustrated on the list below:

Lawn seats	$ .25
Cotton tick	$ .50
Cot	$.75
Bed springs	$ 1.50

According to minutes of the 1927 Fair Board, the owners of amusement rides such as The Wave, which is pictured on page 34, paid twenty-five percent of their daily receipts to the Fair's treasury. The motion picture vendors paid the same percentage. Even the Subordinate Granges, who presented plays, had to return sixty percent of their sales for the privilege.

Baseball games proved to be very popular. Teams from Centre Hall, Boalsburg, Howard, Lock Haven, Milroy and Rebersburg played in a four-day tournament. The final game of the 1927 playoffs netted over eighty-three dollars to the Fair's coffers.

Family, friends and staff at D.C. Keller's Boarding House pose for a picture.

Thrill seekers of the 1920's ride "The Wave," the Fair's first amusement ride which turned to the tunes of "Margie," "Whispering" and "Alice Blue Gown." It was owned and operated by Joseph Thal of Bellefonte. The large open wheel spun like a top while moving up and down and side to side.

In 1928 telephones were installed for public use. A premium book advertisement cost three dollars. Tent rent varied from three to seven dollars depending on the size of the tent. If you brought your own tent, you could rent a space for one dollar.

Francis Harpster Homan 1912.

During the early thirties, a week-long general admission ticket cost just fifty cents. Each ticket had a dated daily ticket that was perforated and surrendered at the gate. Automobile tickets were also fifty cents. In 1931 there were three hundred tents. The Centre County Junior Farmers Building served as dormitory and dining facility for young farmers exhibiting their animals. The Recreation Building has since replaced the Junior Farmers Building.

The disasterous stock market crash and the infamous "Black Friday" of 1929, which resulted in the Great Depression and lasted until World War II, had little impact on the fair. Automobile passes were reduced to twenty-five cents in 1933. They remained that way until 1942. General admission to the fair remained unchanged at fifty cents for a week-long ticket. In spite of the economic hardships the Fair managed to survive.

Sometime during the thirties, the tenting area was expanded to accommodate nearly two hundred additional tents. This was the area between Cherry Tree Row and Woodring Ave., which borders the trailer area. The only shade trees were along the railroad right-of-way. Tenters in this area would have to wait a few years for the newly planted trees to provide shade.

Bertha Sharer, who served as assistant to tent secretary Sarah Grove during the thirties related these memories in a 1977 interview in *The Centre Democrat.*

"In the 1930's the tenters used wagons to do their moving. One of the tenters had a small baby who needed goat's milk, and the mother brought the goat to the Fair. Another time, a man went to his farm in State College to care for his animals and was unaware that his pet hen had jumped into his truck. She rode all the way back to the Fair where they kept her in a crate until they moved back home."

Circa 1920

Did you know that the Fair Board seriously considered installing a nine-hole golf course on the fairground? It was the subject of several months of lively discussion. They even thought of converting the Junior Farmers Building into a Club House.

Agricultural exhibits were well attended by farmers and curiosity seekers.

Above: the De Laval Company exhibits an early cream separator.

Left & Below: Farm tractors from International, Huber and Fordson prove to be a popular gathering place for men and boys.

An early demonstration of the power of the steam driven tractor.

Fair families were never too busy to pose for the photographer at the "town of tents."

The John Wert family of Tusseysink enjoying a picnic in the early 20's. Note that the car seats from vehicles like the 1913 Buick in the background were used for picnic benches.

Shortly after the purchasing of the grounds all Granges set one day aside on which trees were planted next to the railroad where it was believed they would be the most useful.
– Centre Hall History

By 1908 the number of tents had increased and trees began the transformation from wheat stubble to a fairground.

The trees began providing shade for fair goers in the 1920's as porches and other amenities began to appear.

1916 brought a small charge for parking cars but there was no fee for horse drawn vehicles.

In the days of few tents and automobiles, parking in front of your tent was allowed.

Cookstoves stood at the rear of every fourth tent.

Below:
Admission fees required the building of gates. The first gate was completed in 1922.

The Junior Farmers Building above, completed in 1926, had a dining room and dormitory style sleeping quarters for young exhibitors of livestock and agricultural products.

Proud livestock exhibitors stand before the watchful eyes of the judge while spectators look on from the background.

America's favorite spectator sport, baseball, was a popular Fair attraction. Teams from Centre Hall, Rebersburg, Milroy, Boalsburg and Howard played their Championship Tournament at the Fair.

This picture postcard taken from the air gives a view looking south. Notice the new tenting area lacks shade trees and the baseball diamonds in the foreground. The fairgrounds consisted of 70 acres.
Below: A panoramic view of Penn's Valley taken from the top of Mt. Nittany.

The 1940's brought some unusual changes to the Fair. Remember the cannon that graced the front of headquarters? The cannon arrived in Centre Hall in 1897 as a gift from the US government to the local Grand Army of the Republic (GAR) Post. The cannon was a 30-pound Parrot Rifle and was originally mounted on a wooden siege carriage drawn by two horses. It weighed 4,200 pounds, was 131 inches long, had a rifled bore 4.2 inches in diameter and hurled a 29-pound projectile 2,200 yards or 1.25 miles. The cannon was donated as scrap metal during World War II.

The fair was cancelled in 1943 due to rationing of gasoline and other commodities. A one-day picnic was held instead and was well attended according to the newspaper accounts.

1944

GRANGE FAIR
CENTRE HALL, PENNA.
Aug. 26 - Aug. 31, 1944

GENERAL ADMISSION

SHOW YOUR TICKET

To be readmitted to the grounds, present both ticket stub and Pass-out check for the day.

Admission	67c
Federal Tax	13c
Total	80c

The number of tents continued to grow and exceded 550. Weekly admission in 1944 was raised to 80¢ with 13¢ going to Federal Tax. In 1948 the weekly admission was raised to $1.00. Daily admission was 50¢. Entertainment began to change with the introduction of vaudeville acts and Grand Ole Opry shows. This was the beginning of the predominantly country and western entertainment enjoyed today.

President Harry Truman announced the end of World War II on August 15, 1945, one week before the Fair began. Jubilant and sometimes-impromptu celebrations occurred throughout the county. The impact on the Fair was unexpected. All previous attendance records were shattered. The demand for tents by families climbed higher than any time since 1939. Of the 425 tents, 75 were privately owned and the resident fair population was just over 2,100.

Grange Fair Cancelled
According to an announcement made last night by fair officials, the Grange Encampment and Centre County Fair will not be held this year.

The 70th annual fair has been cancelled for 1943 by an unanimous decision of the fair committee feeling that manpower and gasoline shortages were almost insurmountable barriers for holding a successful fair and that it was a patriotic duty to call it off.

Below: Cannon in foreground, the auditorium in the background. The Auditorium is 42 feet by 80 feet and was erected by Luse Planning Mill of Centre Hall. It is believed to be one of the oldest buildings on the fairground.

Trains

Grange Park White Station.

Traveling to and from the Fair in the early years was not always easy or pleasant. Horse-drawn buggies and wagons were the standard means of travel. One of the modern conveniences was the Lewisburg, Centre and Spruce Creek Railroad. Rail service from Spring Mills to Lemont began in 1884. Fair patrons rode the train to Centre Hall and later to Grange Park White Station. During the Fair special trains provided frequent service to the fairgrounds. Automobiles soon became the preferred method of transportation to the Fair and passenger service was discontinued in 1952. Freight service continued until June 1972. The wrath of Hurricane Agnes destroyed most of the railroad's bridges and overflowing streams and floodwater extensively damaged the rail bed. It was declared a total loss and was not rebuilt.

Above: A poster advertising the Fair dates and special Fair train schedule (shown at right).

SPECIAL TRAINS

On September 13th, 14th, 15th, and 16th, as follows:

EASTWARD.			STATIONS.	WESTWARD.			
P.M.	A.M. (This train will be run September 12th also.)	A.M.		A.M.	P.M.	P.M.	P.M.
5:00	10:30	6:55	Lv. . . . . . Bellefonte . . . . . Ar.	9:10	4:25	7:30	10:30
5:07	10:37	7:03	" . . . . . Axeman . . . . . "	9:04	4:19	7:22	9:54
5:12	10:43	7:08	" . . . . Pleasant Gap . . . . "	9:00	4:15	7:18	9:48
5:24	10:55	7:25	" . . . . Dale Summit . . . . "	8:51	4:03	7:06	9:36
5:30	11:00	7:32	" . . . . . . Lemont . . . . . . "	8:46	3:58	7:00	9:30
5:35	11:05	7:39	" . . . . . . Oak Hall . . . . . "	8:42	3:53	6:55	9:25
5:44	11:12	7:47	" . . . . . Linden Hall . . . . . "	8:37	3:47	6:48	9:18
6:00	11:30	8:10	Ar. . .Centre Hall (Grange Park) . Lv.	8:24	3.30	6:30	9:00

Special train leaving Centre Hall at 9:00 P.M. will be run through to Milesburg.

WESTWARD.	STATIONS.	EASTWARD
A.M.	Lv. Ar.	P.M.
8:55	" . . . . . . Sunbury . . . . . . "	9:30
9:05	" . . . Northumberland . . . . "	9:20
9:18	" . . . . . Montandon . . . . . "	9:05
9:28	" . . . . . Lewisburg . . . . . . "	8:55
9:37	" . . . . Beihl . . . . "	8:46
9:42	" . . . . . . Vicksburg . . . . . "	8:40
9:52	" . . . . . Mifflinburg . . . . . "	8:31
9:57	" . . . . Barber . . . . . . "	8:25
10:03	" . . . . Swengel . . . . . "	8:20
10:06	" . . . . . . Millmont . . . . . "	8:17
10:14	Ar. . . . . . Laurelton . . . . . Lv.	8:08

WESTWARD.	STATIONS.	EASTWARD	
A.M.	Lv. Ar.	P.M.	P.M.
10:25	" . . . . Pardee . . . . . . "	7:57	. .
10:37	" . . . . . Cherry Run . . . . "	7:45	. .
10:46	" . . . . Paddy Mountain . . . "	7:35	. .
10:50	" . . . . Ingleby . . . . . . "	7:31	. .
10:57	" . . . . Coburn . . . . . . "	7:23	5:40
11:05	" . . . . Zerby . . . . . . "	7:16	5:30
11:15	" . . . . Rising Spring . . . . "	7:07	5:18
11:23	" . . . . . Penn Cave . . . . . "	6:58	5:10
11:30	Ar. . .Centre Hall (Grange Park) . Lv.	6:50	5:00

The Fair Comes of Age.

After more than seven decades of success and several generations of faithful patrons, the Fair entered a new era. Improvements, new ideas and new construction made their mark. During the fifties and sixties the Fair continued to grow slowly. Tents numbered over 600. Fair goers of the fifties witnessed the first family camper trailers and some converted buses, although some concessionaires brought trailers to the Fair in the late forties. Plays in the auditorium, band concerts and talent shows on the bandstand were still primary sources of entertainment.

Railroad passenger service to the Fair was discontinued in 1952. Freight trains continued to provide daily service and the noon train always provided a bit of extra excitement during the Fair. Before the show arena was built cattle judging took place along the fence next to the railroad. As the train approached Centre Hall, the engineer blew the whistle just as the engine went by the cattle judging area. Pandemonium set in as frightened cattle tried to scatter while their owners hung on for dear life.

A minimal charge for grandstand shows took effect in 1953. New bleacher seating, stage lighting, a wooden fence around the stage area and a ticket booth were added to accommodate a wide range of entertainers and larger audiences. Fair patrons could also get a free chest x-ray and there was a tractor-driving contest for young people.

This was the same year that the two large hog barns burned to the ground on the 28th of August. The story goes that the town folks complained of large number of flies coming from the hog barns. Being good neighbors, the Fair

ground crew sprayed the barns with an oil based fly spray. Allegedly, some children playing with matches accidentally set fire to some straw in one of the barns. By the time it was discovered, it was so intense that it spread to the second hog barn. The conflagration lasted about thirty minutes. Fortunately, no animals or people were hurt. The damage estimate was between ten and fifteen thousand dollars.

The Central District of the Pennsylvania Holstein Breeders staged their first Black and White Show on Saturday, August 28, 1954.

Horseshoe pitching, an ancient past time reaching back to the days of the Roman Empire, was also part of the Fair's agenda. This contest was limited to "any person who obtained his living by farming." The winner would be declared the Centre County Champion and would go on to the State Farm Show for the State competition in January.

The playground was refurbished in 1956 with new swings, sandboxes, seesaws, and merry-go-rounds. A new fence, partially enclosing the area, was added for greater safety. New areas were also added to the tent camp and public parking fields. At the Main Gate, which was originally built to handle horse drawn vehicles, the entrance and exit lanes were marked with stakes to help drivers when traffic was heavy. It became so congested that the Fair Board decided to add another gate. The new gate, which is now Gate #1, was constructed in 1957 and provided much easier access for the heavy volume of traffic. The Main Gate is now Gate #2.

August 18, 1960, *Millheim Journal*
The Grange Fair Board purchased 58 new tents to replace those worn and damaged. The new wing of the Headquarters Building was completed and the original portion rearranged to give more adequate facilities for the staff that care for campers and visitors inquiries.

Those observing the annual Veterans Day program for 1959 were treated to a mock warfare and tank demonstration conducted in the parking fields. Troop L, 3rd Reconnaissance Squadron of the 104th Armored Cavalry, Pennsylvania Army National Guard put on the show. It was followed by a parade of veteran's units led by the Gardner Guards drill and marching unit.

Dairy cattle being judged prior to the building of show arena.

A show arena was constructed in the sixties and was reconstructed along with the beef barns in the late sixties when a heavy snow collapsed the roofs. The poultry barn was converted into a concession area when poultry exhibits were banned due to avian influenza. This building later became the goat barn. Recreational vehicles made their fairground debut in the early sixties and have continued to increase in numbers as well as size and popularity.

When recreational vehicle owners groups like the Penn Coachmen and others like the Good Sam Club were first organized, the fairground proved to be a good meeting location. This required some operational changes for the Fair Board and a new source of revenue.

The parking fields, which were not used except during the Fair, were rented to local farmers to graze young cattle or grow grain. Cattle were removed a month prior to the Fair so that most of the evidence of their having been there would be gone by Fair time. As the requests from recreational vehicle groups became more frequent, it became necessary to remove permanently the grazing herd.

"Grangers should stay out of politics. Farmers, do your duty and try to go to Heaven".

Former Governor Andrew G. Curtin

In 1968 the tenting area was expanded to accommodate 700 tents. Camper trailers and recreational vehicles grew to 200 and had to be moved to the far west side of the fairgrounds.

Governor Milton Shapp was the guest speaker in 1966. He addressed what was believed to be one of the largest audiences ever.

Political rallies were a major part of the Fair for decades. Political speeches were a part of the program since the first Picnic. Former Governor Andrew G. Curtin, who was Pennsylvania's Governor during the Civil War, was a popular speaker during the early days of the picnic. On one such occasion he was reported to have said, "Grangers should stay out of politics. Farmers do your duty and try to go to Heaven." Twenty thousand people heard him say that.

Governor Curtin's advice went unheeded. As the crowds at political rallies got larger, the congestion around the bandstand made moving about nearly impossible. Some members of the Fair Board thought the rallies became disruptive to the point that others could not enjoy the fair. The Board's efforts to move the rallies to the grandstand were ignored. Political leaders liked the congestion because it put the candidates "shoulder to shoulder" in the audience. Moving to the grandstand would place the crowd in the bleachers separating them from the stage and the political candidates. Democratic County Chairman, Mrs. Marie Garner commented, "It's too bad. I think it's a thing people looked forward to." She added that the Fair offered the only close touch there is during the year between members of the two major parties. "I think it made for more congenial relations between the parties," she concluded. Regardless, the Fair Board prevailed, and 1967 marked the last year of political rallies. Politicians still come to the Fair. Instead of the bandstand, they use political party tents located on the midway. They campaign among the crowds in the exhibit buildings, on the midway, visit the Fair residents, and ride in the parade.

New Horse Barn April 1999.

There were more improvements made in the seventies. The Grange Exhibit Building was constructed and a new aluminum grandstand and sound stage completed. Individual tent owners were asked to rent a tent from the Fair. Odd sized tents were replaced with 14 x 14 wall tents. Aging wooden tent floors were replaced with asphalt.

The Centennial Celebration was held in 1974. By now the tents exceeded 700 with 425 trailers and recreational vehicles. The resident Fair population exceeded 6,000.

Wooden tent floors of the sixties.

Begun in 1995, the tent decorating contest is just one of the many Fair activities.

Trailers and recreational vehicles continue to come in all shapes and sizes.

The newly refurbished playground is a popular attraction for the younger crowd.

The Fair Board purchased the adjoining 212-acre Obert Ilgen farm in 1984. This provided the room for expansion and many improvements. The immediate additions for the 1985 Fair included the new horse show arenas, expansion of the RV and camper sites, and the Gate #4 entrance on Homan Lane from Route 45. The midway was also expanded to accommodate more concessions with the construction of the Triangle and the Blue exhibit buildings.

From 1986 to 1990, a tractor pull area was established and the new midway further expanded to provide space for more concessions. In 1990, the trams were introduced to provide transportation to and from the parking lots, to the grandstand, various entrances to the midway, and animal exhibit barns.

Underground electric service to the tenting area began in 1992 and was completed in 1996. Showers were also installed in 1994. A new dairy barn, measuring 72 feet wide and 218 feet long, was built in 1994 replacing the old barn.

Expansion of the new midway area continued in 1995 with roads, lighting and utilities to accommodate concessions. More roads and walkways were improved with a coating of asphalt. The new Administration Building was built in 1996. Teams of Emergency Medical Technicians replaced the long familiar first aid tent next to the boarding house. New lighting in the parking lots was completed in 1997.

To date, the year with the most improvements was 1998. New electric service was brought in from Route 45. Gate #4 was removed and a new, relocated, four-lane gate was constructed to handle a larger volume of traffic.

Many minor improvements were made to make the Fairgrounds more accessible to disabled persons. The tractor pull area was relocated south of its former location to make more room in the new midway. The old Shaffer Electric Building was replaced with a new metal building. Jubilee Grove was constructed in front of Headquarters with plantings, lighted flagpoles, benches and engravable flagstones for memorials. New street signs were installed and plantings of flowers and shrubs added a new dimension of beauty. A new horseshoe pitching area was constructed near gate #4. The playground was refurbished with new equipment and a thick mulch surface, for the safe enjoyment of the younger crowd. And finally, the Fair's long awaited first connection to the sewage system took place.

Articles about the Fair have appeared in *The National Geographic, New York Times, USA Today* and *The Philadelphia Inquirer.* A website, www.pafairs.org/centrecountyfair was established in 1998 as another way to promote the Fair. It is hyperlinked to the Centre County Convention and Visitors Bureau. Hand canceling of postage stamps at the Fair began in 1996. This year's cancellation stamp includes the 125th Anniversary logo.

There were two major improvements in 1999. The first, the installation of flush commodes in the restroom next to the cattle barns. Second, the construction of a 108-stall horse barn.

The next sections highlight various aspects of the Fair. They include Agriculture, Parades, Entertainment, Concessions and Exhibits, Rides, Souvenirs, and the Fair Board.

Aerial photograph 1998.

"Come on we need to get to the show."

August 26, 1954, *Millheim Journal*
For the first time in history Grange Fair dairy cattle will be on exhibit over the weekend. The First District, Central Pennsylvania Holstein Breeders will stage their first black & white show at the fair on Saturday, August 28.

Agriculture has always been the primary focus of the Fair. Farmers, Granges, 4-H, Future Farmers of America, and individuals exhibit their best as they compete for ribbons and premiums. The wide array of competition ranges from livestock, grains, forage grasses, fruit and vegetables to clothing, quilts, crafts, horticulture and baked goods. Educational exhibits provide a wide variety of information awareness and assistance with problem solving. Commercial exhibits of farm machinery play a significant role in keeping the agricultural community informed about the latest advances in technology and farming methods.

The first year that farm machinery was exhibited was in 1887, the last year the Picnic was held on Mt. Nittany. As farm machinery became more sophisticated, the newly developed internal combustion engine was used to power tractors instead of steam. Two of the area's early dealers were David Bradford of Centre Hall, who sold International tractors and Cleveland Eungard of Spring Mills, a new dealer for LaCrosse Happy Farmer tractors. Happy Farmer tractors were first exhibited in 1919. The story goes that Cleve received a special pair of coveralls embroidered with the Happy Farmer logo just for wearing at the Fair exhibit. His young son, Kenneth, who was about four years old at the time, had to have a pair of coveralls just like his dad's. Ken's mother bought him a new pair of coveralls and embroidered them with the Happy Farmer logo so he could look just like his dad. It was a sight to behold the "twins" at the Happy Farmer exhibit.

The exhibits continued until about 1930. The Depression stopped all the machinery exhibits until 1935. A number of farm implement dealers returned to the Fair, but they only stayed until 1939. A disagreement between the dealers and the Fair Board over the exhibit fee led to a boycott. The hiatus lasted until 1950. An agreement was worked out where the Centre County Farm Implement Dealers Association rented the space between the north and south gates (between gate #1 and gate #2). The Dealers Association then allotted space to individual dealers based on their needs. Over the next several years the number of implement dealers decreased. The space they once occupied slowly gave way to the demand for more trailers and recreational vehicles. After the Ilgen property was purchased, farm implement dealers returned to occupy space on the new midway.

The Open and Junior Livestock Shows have become an integral part of the Fair. Shows include beef cattle, dairy cattle, hogs and sheep. In the late 1980's the Dairy Goat Show began and Market Goat Shows were added in 1990. There has been a steady increase in the number of animals shown. In the early 1960's as few as five steers were shown. Now there are over sixty steers and twenty breeding animals. All cattle must be dehorned and all animals must meet State health requirements prior to being registered into the Fair.

The Junior Livestock Sale started in 1972, when Ken Dietrich, FFA Advisor in the Bellefonte School District, initiated the idea of having a sale during the Fair. In years prior, hogs and sheep were taken to the Centre Hall Livestock Auction during the Fair. Steers were sold privately or taken to other sale locations such as Pittsburgh. E.M. "Jack" Smith "cried" the first auction from the back of his pickup truck with Don Dreibelbis assisting. Since then, the number of animals sold and the support provided by local businesses, professionals and

Grange Exhibits

Judge Les Burdette - Early 1950's.

Eugene Wolfe lends a firm but gentle hand.

individuals has steadily increased. The 1972 sale records show that 85 animals were sold. Clerks for the early sales included Jack Guisewhite of the Millheim Farmers Bank and Bill Rockey of the Spring Mills Bank. In 1980, the Centre County Junior Livestock Committee and Penn State Cooperative Extension began managing the sale. The First National Bank of Centre Hall, now Northwest Savings Bank, provided staff to collect monies and distribute checks.

"The sweet taste of victory" - Bunnies nibbled their prize ribbon.

Picture perfect.

Sale records from 1981 indicate that 218 animals were sold for a total exceeding $77,000. In 1998, sale records show that 268 animals were sold for more than $120,000.

Draft horses were exhibited in the early days of the Fair. As tractors and farm machinery were developed, farmers slowly replaced horses with tractors. The close of World War II brought about a dramatic change in farming methods as the tractor made its debut. Cumbersome horse drawn implements gave way to modern, efficient equipment, pulled by powerful tractors. Draft horse exhibits slowly disappeared and were gone completely by the early fifties. During the sixties and seventies, the number of draft horse breeders all but disappeared.

Renewed interest in draft horses generated a dozen or so breeders in the county. In 1981 draft horses made a comeback to the Fair's schedule of events. Ten county breeders with 25 horses entered the competition in two classes. There was a halter class in which horses are judged on their conformation, gait and general good looks. In a second class, teams of two horses with brass and chrome show harnesses were hitched to wagons and drivers demonstrated their ability to maneuver. A standing room only crowd was treated to a show featuring three of the five nation's breeds — Clydesdale, Percheron and Belgian.

The Saturday morning shows were conducted at the grandstand until 1984. They were relocated to the new horse arenas in 1985 and have been a big attraction ever since. The new horse arenas are also the location of the Keystone Central Horse Show, that takes place over the weekend of the Fair. A new 108-stall horse barn was completed this year (1999) to accommodate the Fair shows and other horse shows.

Parades

Parades have always been an integral part of the Fair. There was even a parade at the first Picnic composed of Grangers, dignitaries, and local bands. Audiences enjoyed the veterans parade until the mid-thirties. The parade became smaller as the number of war veterans decreased.

Grange floats first appeared in 1939, and have been featured ever since. Today, the parade has a wide representation of community organizations with 4-H, FFA, fire companies, twirling groups, politicians, local high school bands, women's clubs, senior citizens, girl and boy scouts, cub and brownie scouts and local veteran's groups.

PROGRESS

CONSIDER THE LILIES
OF THE FIELD
HOW THEY GROW

ATOM POWERED

SPRING MILLS 4HCLUB
WORK
Makes the Best Better

PaTrons of Husbandry

Entertainment

Entertainment at the Fair has varied over the years. The early picnics featured prominent speakers like Governor Curtin and concerts by local bands. After the Fair moved to its permanent home, other forms of entertainment began to appear. Motion pictures were shown in front of Headquarters. Guest speakers and plays presented by Subordinate granges and the Young Patrons of Husbandry (Y. P. of H.) filled the auditorium. Band concerts and political speeches were given at the bandstand. Special events, like the oxen tug-of-war featured in 1939, had four teams of trained oxen with pulling demonstrations and stunts. Other animal shows were Professor Keller's "Jungle Killers" featured in 1940. Circus acts included aerial performers and acrobats. In 1949, "new" entertainment made its debut with a black face comedy team from New York called "Jackson and Miss Sugar" and American master magician, Josef Smiley with his company of "Wonder Workers" who presented fifty thrilling illusions in a magic act. The entertainment slate for 1955 included "Frontier Days," a novelty circus program with horses, cowgirls, cowboys and Indians, with roping, jumping, shooting and riding acts. This was the first year that stars Martha Carson and Jimmy Dickens from Nashville's Grand Ole Opry made an appearance.

In 1973, a new stage and grandstand was constructed to accommodate the growing crowds. Aerial circus acts performed three-to-five nights with fireworks as the grand finale. Scheduling different

Patty Loveless
1989

Boxcar Willie
1982, 1983

The perennial race for the best seat in the house.

Jerry Reed 1988

Louise Mandrell 1989

Holly Dunn 1989

Tom Wopat 1990

Tom T. Hall 1986

Chubby Checker
1981, 1988, 1996

1987

Tracy Lawrence 1992

1975

This 1965 postcard depicts the Bandstand which was located at the site of the present Jubilee Grove.

David Kersh 1998

Paul Revere & The Raiders 1998

nightly entertainment, with virtually unknown performers, evolved during the mid-seventies. The stage was enclosed, along with the addition of lighting and a sound system. Who would have thought that names like Grandpa Jones, Porter Wagoner, Myron Floren, Tom T. Hall, Crystal Gale, Jeannie Riley, Chubby Checker and Box Car Willie, just to name a few, would be among the nation's top entertainment personalities? Then there's Garth Brooks, David Kirsch and The Kinleys who lead the country and western entertainment circles. They also got their start in the entertainment world performing the summer fair circuit. Now there a two shows each night to accommodate the crowds of enthusiastic fans.

Garth Brooks 1990

Bobby Burgess and Elaine Niverson of the Lawrence Welk Show.

Sylvia 1987

Concessions & Exhibits

Exhibits and concessions first appeared in 1877, when the Fair was held on Mount Nittany (Centre Hall Mountain). They consisted primarily of farm machinery, with a very small number of food and beverage vendors. After the Fair moved to its permanent home in 1890, pictures of the concessions show the "stick stands" which consisted of a wooden framework that was easy to assemble and disassemble. Some added a canvas top or a cloth cover to lend some protection from the elements. Commercial exhibits consisted of the latest farm machinery and other labor saving devices, like the tractor. During the 1920's and well into the 1930's, automobiles were exhibited by local dealerships. Household appliances like washing machines, sewing machines and wood and coal fired cook stoves were favorite exhibits among the women. As attendees came from farther distances, it was difficult to bring all the food that they needed. Several local merchants set up grocery stores to provide fresh baked goods, meat, fruit and vegetables. As more people visited the Fair for the day, the need for food and beverage concessions also grew.

The Boarding House met the needs of many Fair patrons. The building was built in the early 1900's. At first, local Granges prepared and served meals. In 1932, Lyman and Helen Torrey of Boalsburg acquired the Boarding House from the Hassinger family. All of the family was pressed into service. Helen baked pies long into the night. The majority of the food came from their Boalsburg farm with the rest coming from local suppliers. They used linen table clothes, china dishes, and silverware. The family stayed in two large tents. They carried this tradition until 1980 when age and poor health took its toll. Phil and Marsha Poorman are the present managers of the Boarding House.

"Around 1900 Jasper Spigelmeyer & Samuel Young got on a train at Mifflinburg and came to the Fair, made lemonade and sold it for $.05 for all you could drink."

Other food concessions had limited but specialized menus because of their size. When automobiles became more affordable and roads constructed to replace cow paths and trails, Fair concessionaires began attending several events and the fair circuit was born along with a multi-million dollar business. Today there is a wide variety of foods available, representing many cultures.

Games of chance also increased in numbers and variety. Dart games, duck ponds, baseball pitch, shooting galleries, casino games, bingo, and the penny pitch barely scratch the long list. Showing off your skill also had its reward in the form of prizes. Stuffed animals, canes, leis, and gold fish lead the list. One particular game of chance has an unusual reward. You can't taste it and you can't take it with you. "The Striker," one of the Fair's early games, challenges your ability to swing a wooden mallet with enough force to cause a weight to slide up a twenty foot high metal track and ring a bell..."Bong"...your reward. Owned by Jerry and Lisa Bartlebaugh, "The Striker" is still located at its original site.

"Five a cake, five a cake, up a hill, down a lake the more you eat, the more I'll make, 5 a cake, 5 a cake."

Refrain of ice cream sandwich man at Grange Fair in 30's and 40's.

Ague's 1950's ice cream stand was built on a trailer chassis. The sides folded down so it could be towed from fair to fair.

Rides

Jack Garbrick and a helper wire electric lines used to power rides. In earlier years rides were powered by gasoline engines.

Amusement rides have always been a Fair favorite. The first Fair ride was a contraption called "The Wave." Pictured on page 34, it was introduced to the Fair around 1915 and was a fixture for many years. Owned and operated by Joseph Thal of Bellefonte, "The Wave" was an open wheel with seats facing toward the outside. There were four sets of steps by which to enter and leave. While it turned in a circle, it also raised up and down and moved side to side to the tunes of "Margie," "Whispering," and "Alice Blue Gown."

From the early 1930's to 1954, the Reithoffers of Wilkes-Barre, PA provided rides. From 1955 to present, Centre Hall's own amusement ride company, Garbrick Manufacturing, has provided Fairgoers with a variety of rides. In 1955 a merry-go-round and a Ferris wheel were purchased from Hecla Park and brought to the Fair along with eight or nine kiddie rides. Tickets were 12 for one dollar.

Rides then were located just south of the midway where Pomona Street is now. Eventually, they were moved to their present location because of the demand for concession space.

According to Jack Garbrick, the most popular rides for the past 44 years have been the merry-go-round and the Ferris wheel. Countless children have had their first merry-go-round ride at the Fair with parents or grandparents holding them tight when the horses leap into action. There are now twenty rides at the Fair, staffed by thirty employees.

Garbrick Manufacturing is known world-wide for the quality and reliability of their rides. They invented and patented the Merry Mixer and the folding Ferris wheel. They also built the first all metal merry-go-round with metal horses, manufactured in Leavenworth, Kansas.

Airplane rides have always been popular. Among the early aircraft was a Pitcairn Autogyro. It was propeller driven but had rotor blades similar to a helicopter for wings. The pilot used the field where the RV's now park for a landing strip. Sherm Lutz also provided rides for a number of years. In 1965, Jack Garbrick began flying his Cessna 172, using Obert Ilgen's field for a landing strip. Eugene Tate, who was also Garbrick's flight instructor, served as a pilot. "Aviators" had to climb a set of steps to go over the fence before a gate was finally installed. Flights to "See the Fair from the Air" lasted 6-7 minutes and provided more than 80% of the riders with their very first airplane ride. 1984 was the last year for plane rides. New Federal Aviation Administration Regulations prohibited approval of a temporary airstrip on the fairgrounds.

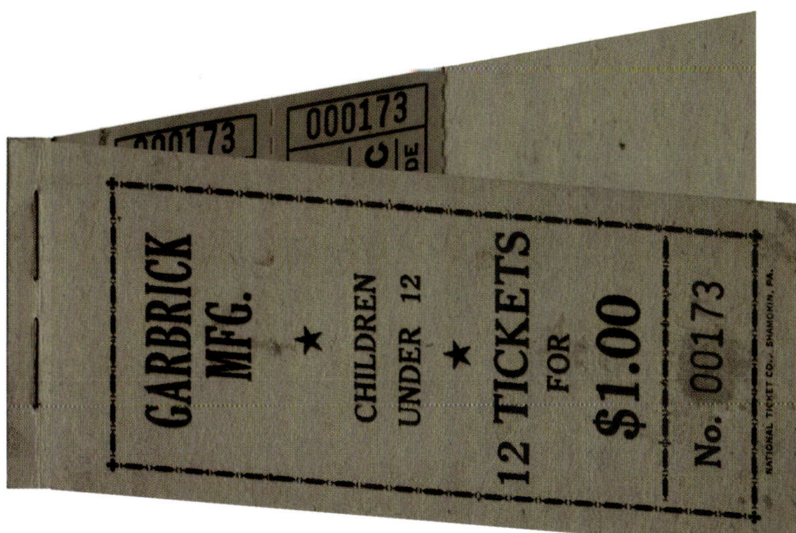

These postcards were early favorites.

Souv
1908
Grangers' Picnic

Souvenir Ware

Red Ruby Glass

Consisting of custard glass, pattern glass (in emerald green, ruby stained, and apple green), and china, these mementoes were popular from the 1890's to the 1920's. Usually consisting of household items — cups, pitchers, bowls, toothpick holders, tumblers, salt & pepper shakers, and plates, the price was often less than a dollar. The article was sold to the local vendor in blank, and then decorated at the event with etching, painting or stenciling. They were frequently taken home and placed in the china closet or in the hope chest.

100th Year

Custard Glass

Milk Glass Hair Receiver & China

Red Ruby Glass

Red Ruby Glass

China

Milk Glass

Custard Glass

Candle

Custard Glass

Vasoline Glass Pot

Paperweight

Red Ruby Glass

Paper Weight

Tiny Iron Pans

Emerald Glass

Vasoline Glass

Milk Glass

Raffle Tickets

Match Safe

Milk Glass, Emerald Glass & Vasoline Glass

Emerald Glass

Blue Speckled & Apple Green Glass

Purse Mirror

Custard Glass

Red Ruby Glass

Plate from 1974 Centennial.

China

Pennant & Bumper Sticker from 1974 Centennial.

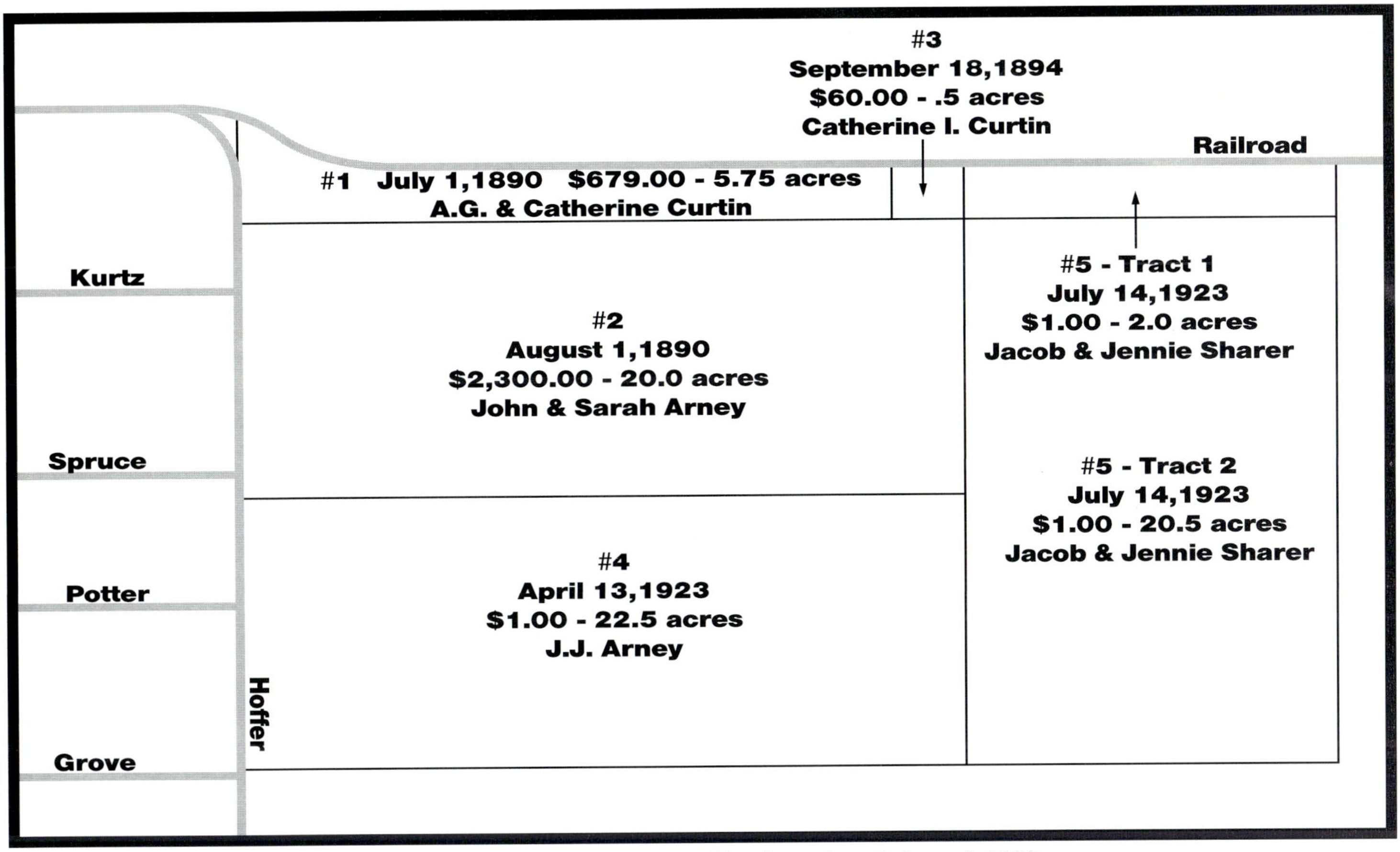

#3
September 18,1894
$60.00 - .5 acres
Catherine I. Curtin

Railroad

#1 July 1,1890 $679.00 - 5.75 acres
A.G. & Catherine Curtin

Kurtz

#2
August 1,1890
$2,300.00 - 20.0 acres
John & Sarah Arney

#5 - Tract 1
July 14,1923
$1.00 - 2.0 acres
Jacob & Jennie Sharer

Spruce

#5 - Tract 2
July 14,1923
$1.00 - 20.5 acres
Jacob & Jennie Sharer

Potter

#4
April 13,1923
$1.00 - 22.5 acres
J.J. Arney

Hoffer

Grove

This map illustrates the tracts of land purchased through 1923.

The dotted line outlines the Obert Ilgen Farm which was purchased in 1984.

Show Arena

Tractor Pull

The Opening Ceremony of the 1998 Fair which included the Dedication of Jubilee Grove.

Big Blue and the Triangle Building 1985.

New Administration Building 1996.

Behind the Scenes

The Fair Board currently consists of 38 members representing their respective Granges, Young Patrons of Husbandry, 4-H, and FFA. Working year round, this small corps of volunteers ensures that the annual fair is a wholesome, educational and family-oriented event. To accomplish this monumental task requires an organization that can develop and implement plans, initiate new ideas, and generally look for ways to improve the Fair. Each board member serves on several committees. Some members have decades of service on the Board. Their expertise and experience is the "magic" behind the scenes. The Fair Board deserves everyone's praise for their dedication and hard work.

Early 1990's Fair Board.

100th Year (1974) Fair Board.

1999 Fair Board.

The Fair has a bright future. The Planning Committee has had many meetings and discussions to plan improvements for the next several years. Among those are the following:

- Replacing all water lines.
- Flush toilets in all restrooms.
- Construction of an additional playground.
- Construction of concession buildings.
- Construction of a 400 seat banquet hall.
- Asphalting streets and walkways.
- Enlarging the trailer and recreational vehicle area.
- Construction of two little league baseball fields.

It is obvious that this book is not all-inclusive. There are many stories that could not be included because of space constraints. The primary focus was to note the significant historical events and their impact on the Fair. What is most significant is that Leonard Rhone left us a legacy and a tradition that has lived on because of his exemplary leadership. It is therefore fitting to close this volume with the poem entitled "Granger's Fair," written in 1936 by Harvey Flink, Centre Hall's poet laureate. The poem captures the very essence of what the Fair has been, still is and will always be.

Granger's Fair

All hail, to Granger's Picnic,
The chief of all events!
This week the crowded highways
Lead to the town of tents.
All hail, to Granger's Picnic
And stirring incidents!

The exhibition building,
With its four spacious wings,
Its tale of wealth and culture
To visitors now brings;
For here are garden products,
Fine fruits and fancy things.

The barns are interesting–
The poultry house is, too–
With horses, sheep and cattle–
With laying hens to view;
And automobile dealers
Have cars to show to you.

The crowds are surging forward,
Like waves in restless seas;
Some pausing where the band plays
beneath the maple trees,
And some go to the ball ground,
Where games are sure to please.

At night the revels center
About the lighted stands;
The daring seek amusements
Of many kinds and brands,
While happy lovers loiter
And hold each other's hands.

At last the crowd grows thinner,
Less lively is the fun.
The Ferris Wheel is idle,
The Wave has ceased to run;
The cars are starting homeward–
Another day is done.

And after all is silent,
Except the cricket's drone.
A happy spirit wanders
About the grounds, alone -
The spirit of the founder,
The kindly Leonard Rhone.

Contributions

The History Book Committee thanks the individuals listed below who contributed photographs, newspaper articles, souvenirs, stories, and artifacts. We apologize for any errors or omissions.

Joyce Adgate - Centre County Library & Central PA District Library - Bellefonte, PA

Richard Ague - Sugarloaf, PA

Pat Armagast - Penn State University

Ellen Berkebile - Lock Haven, PA

Carol Billette - Bellefonte, PA

Jean Bonchack - Centre Hall, PA

Rosella (Stover) Boone - Aaronsburg, PA

Joan Brower - State College, PA

Dick Brown - Bellefonte, PA

Fred & Miriam Burris - Centre Hall, PA

Centre Hall Area Library - Centre Hall, PA

Darlene Confer - Centre Hall, PA

Isabel B. Corl - State College, PA

Jean Corman - State College, PA

Dorothy Delaney - Centre Hall, PA

George & Donna Dickerson - Spring Mills, PA

Sue Foster - State College, PA

Kate Frazier - Pleasant Gap, PA

Mike & Carol Frazier - Pleasant Gap, PA

Daniel Frome - State College, PA

Fred & Dorothy Fry - State College, PA

Robert & Rose Fry - State College, PA

Hal Green - State College, PA

Ben & Maryann Haagen - Howard, PA

Jane Haagen - Bellefonte, PA

Lisa Hawn - Petersburg, PA

Priscilla Rossman Hall - Jersey Shore, PA

Dick Harpster - Centre Hall, PA

Shirley Heidrich - State College, PA

Phyllis High - Boalsburg, PA

Vonnie Henninger - Rebersburg, PA

Louise Horner - Boalsburg, PA

Ellie Huett - Melbourne, FL

Eloise Ingram - State College, PA

Dr. & Mrs. H. Richard Ishler - State College, PA

Roxie Ishler - Bellefonte, PA

Marvin Bair - Madisonburg, PA

Scott Frazier - Centre Hall, PA

Frances Keller - Boalsburg, PA

Donita Rudy Koval - Centre Hall, PA

Aida Lamey - Rebersburg, PA

Norm & Kathy Lathbury - Centre Hall, PA

Mark Lucas - Blanchard, PA

Shirley Luse - Centre Hall, PA

Judith McKinney Mathes - Dayton, OH

Jeff McClellan - Coburn, PA

Janet Biddle McClellan - Julian, PA

Martha McCrain - Howard, PA

John W. McKinney, Jr. - Beaver Creek, OH

Kathy Mothersbaugh - Sping Mills, PA

Clarence Musser Family - Spring Mills, PA

Joyce Neff - Tusseyville, PA

Lee & Judy Pressler - Port Matilda, PA

Progress Grange - Centre Hall, PA

Charles & Barbara Richard - Petersburg, PA

Jim & Janet Rider - Gatesburg, PA

Dean C. Rishel - Spring Mills, PA

Galene M.C. Schock - Cogan Station, PA

Ruth Shreckengast - Millheim, PA

Willard Smith - Madisonburg, PA

Charles Spigelmeyer - Centre Hall, PA

Fred Strouse - Centre Hall, PA

Robert M. Struble - State College, PA

Maryann Taylor - Spring Mills, PA

Bruce Teeple - Aaronsburg, PA

Donald Tomco - Bellefonte, PA

Mr. & Mrs. Donald J. Tressler - State College, PA

Vicki Vallimont - Lewistown, PA

J. Ivan Wasson - Coburn, PA

Bob Weber - Centre Hall, PA

John Wert, Jr. - Centre Hall, PA

Ruth R. Wolfe - Washington D.C.

Joyce Wolfe - Centre Hall, PA

Shirley Walker - Boalsburg, PA

Dave White - Baltimore, MD

Joseph Wojtaszek - Spring Mills, PA

Jack & Tress Yearick - Centre Hall, PA

LeDon Young - Centre Hall, PA

David Ziegler - Centre Hall, PA

SPONSORS

The History Book Committee deeply appreciates the financial support provided by the sponsors listed below, who helped make this project a success.

Kenneth Babe - State College, PA
John Bonfatto - Bellefonte, PA
James R. & Mona Bowmaster - Bellefonte, PA
Cellular One - State College, PA
Richard P. Creasy - Bloomsburg, PA
Forms Plus - Scranton, PA
Dorothy Hartle - Bellefonte, PA
Joseph & Gladys Hartle - Bellefonte, PA
Jay & Mary Houser - Spring Mills, PA
Dorothy Jodon - Bellefonte, PA
Karen Jones - Arlington, VA
Karen L. Jones & John P. Mackie - Arlington, VA
Marlin R. & Betty Koch - Levittown, PA
Marion Grange - Bellefonte, PA
James & Janet McClellan - Julian, PA
Richard G. & Shirley McKelvey - Centre Hall, PA
John W. McKinney - Beaver Creek, OH
Roger L. Myers (Roger's Taco) - Avis, PA
Daniel L. & Carol Royer - Woodland, PA
Thomas N. & Denice Royer-Hewlett - Upper St. Clair, PA
Dean C. & Jean Rishel - Spring Mills, PA
Salem Motors - Selinsgrove, PA
Norman T. & Sandra Seibert - Virginia Beach, VA
Jean Slear - State College, PA
Victor Grange - Boalsburg, PA
Ruth R. Wolf - Washington D.C.

Memorial Contributions:

In Memory of Don & Iola Fry
by the Fry & Heidrich Families,
State College, PA

In Memory of Fred E. Luse
by A. Robert & Inez Luse
Centre Hall, PA

In Memory of Clarence N. Peters
By Anna M. Peters
PA Furnace, PA

In Memory of Eugene R. Wolfe
By his Family
Centre Hall, PA

1999 Grange Fairgrounds

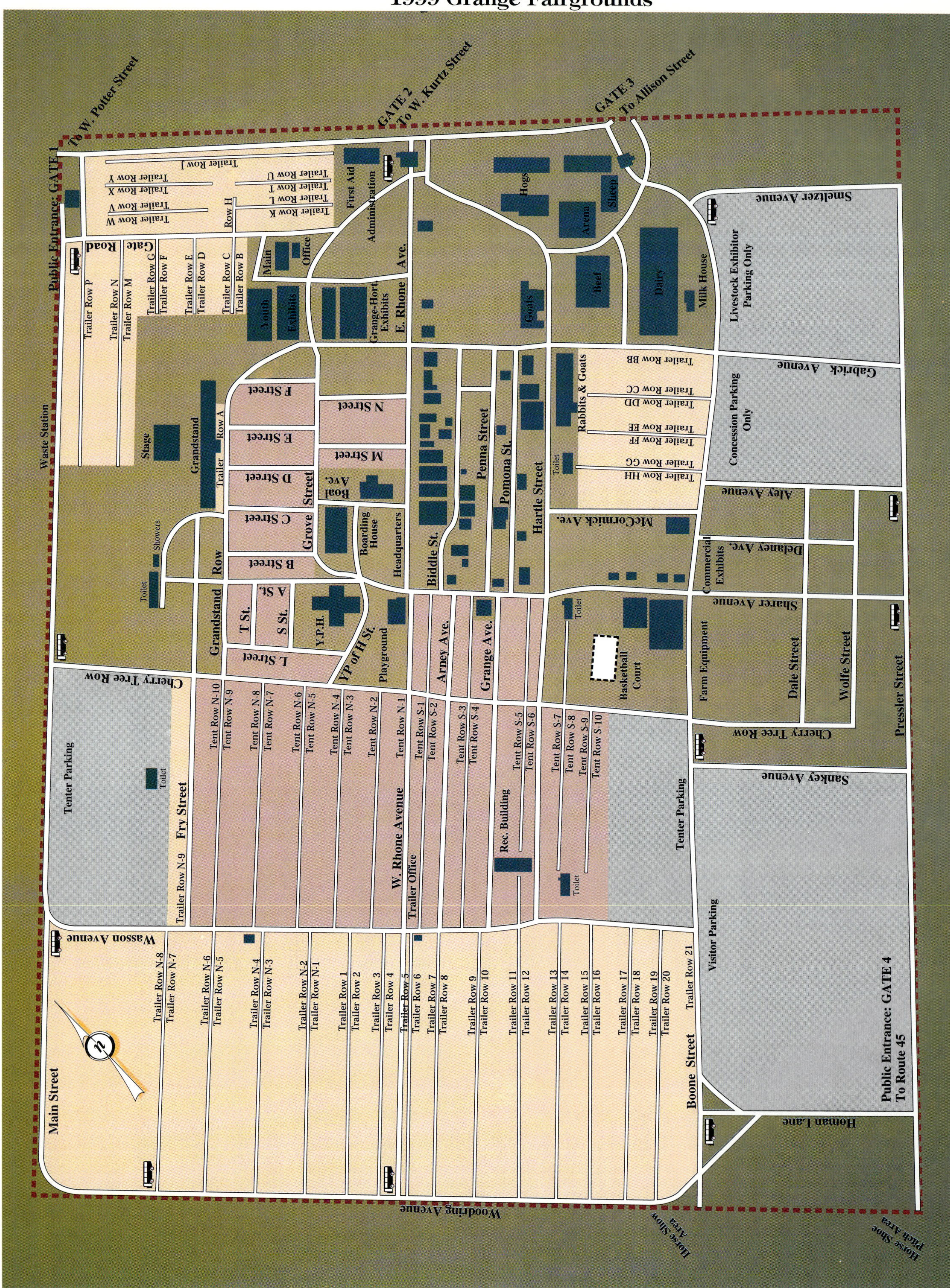